GROUND WATER POLLUTION
AN INTRODUCTION

GROUND WATER POLLUTION
AN INTRODUCTION

Edited by

Dr. M.S. Waghmare and
Dr. Kailash Malode

Delve
PUBLISHING

www.delvepublishing.com

Ground Water Pollution: An Introduction

Dr. M.S. Waghmare and Dr. Kailash Malode

Delve Publishing
4164 Lakeshore Road
Burlington, ON L7L 1A4
Canada
www.delvepublishing.com

Email: orders@arclereducation.com

ISBN: 978-1-77956-358-3 (Hardcover)

Delve Publishing publishes wide variety of books and eBooks. For more information about Delve Publishing and its products, visit our website at www.delvepublishing.com.

ABOUT THE EDITORS

Dr. Mahesh S. Waghmare (b. 1984) is presently serving as Assistant Professor, Department of Soil Science and Agricultural Chemistry, College of Agriculture, Dharashiv, Vasantrao Naik Marathwada Krushi Vidyapeeth, Parhani, Maharashtra. He did his B. Sc.(Agri.) in 2005 and M. Sc. (SSAC) in2007 from College of Agriculture, Latur, Ph. D. (Soils) in 2011 from College of Agriculture, Parbhani (MS). He started his career as a Research Associate, under NICRA (AGROMET) project at VNMKV, Parbhani in the year 2012-2014. He worked as a Subject Matter Specialist (Soil Science) at MGM, KVK, Gandheli Aurangabad (MS) during 2014-2016. He joined as a Assistant Professor in july 2016 at College of Agriculture, Dharashiv (VNMKV, Parbhani).His field of specialization is Soil Fertility and nutrient management, Nano-fertilizer, Soil survey etc. During his total 12 year's tenure of services, under his guidance 05 students completed M.Sc., and he acted as co-guide for 20 students in Soil Science discipline. He has published 02 practical manuals for under graduation, 30 research papers in international and national reputed journals. He had participated and presented research papers more than 10 International, National, State Seminar, workshops and symposiums. Besides these activities Radio talks (10), guest Lectures in scientist training & farmer training programmes (25). In addition to above he has written more than 25 popular articles in reputed regional languageperiodicals and daily news papers. He is recipient of Fellowship from International Phosphorous and Potash Research Institute, Canada (India Programme) during his Ph.D. He received Commendation Certificate for Best doctoral Research Presentation Award from Indian Society of Soil Science, New Delhi in 2012. He is recipient of Young Scientist award for best presentation. He also awarded as Young scientist award from PLANTICA society, Dehradun.

Dr. Kailash Malode (1980) is presently serving as Assistant Professor, Department of Soil Science and Agriculture Chemistry, college of Agriculture, Parul University, Vadodara, Gujarat. He obtained his B. Sc. (Ag.) in 2006 from Marathwada Agriculture University, Parbhani. M. Sc. (Ag.)2009 in Dr. Panjabrao Deshmukh Krishi Vidyapeeth,Akola and Ph. D. Soil Science And Agriculture Chemistry in 2014 from Vasantrao Naik Marathwada Krishi Vidhyapeeth, Parbhani. He started his career as Assistant professor in 2014. His field of specialization is soil fertility, Remote Sensing and Organic Farming. He has published 02 books and 15 research paper in National and international journals of reputed. He has participated many National, State Seminar and symposiums. He has written 02 practical manuals for under graduate course for the benefit of students besides several Radio talk, Lectures in farmer traning programme. Dr. Kailash has also written 21 popular articles in Shati Bharti, Shati pragati and other reputed Magazines. He has Best writer award on 2015 for Krishi dooth popular Magazines of Maharashtra and One Young scientist award in 2020.

TABLE OF CONTENTS

LIST OF FIGURES

ABSTRACT

Groundwater geochemistry studies the chemical composition and quality of groundwater, focusing on contamination from human activities like industrial discharge and agricultural runoff. With freshwater scarcity being a global concern, it's essential to understand the geochemical processes that lead to pollution, including the interactions between contaminants and geological materials. Analysis of water samples reveals the presence of heavy metals and nitrates, and geochemical modelling helps assess pollutant transport and degradation pathways. Poor waste management and excessive agricultural inputs have degraded shallow aquifers, necessitating effective management strategies to mitigate contamination. The book emphasizes the importance of isotopic analysis, mineral interactions, and climate impacts on groundwater dynamics, highlighting the complex interplay within the hydrological cycle and its implications for environmental management. Pesticides in soil interact with microbial activity and soil properties, affecting their persistence and potential groundwater contamination, necessitating careful management and optimized application strategies. Advancements in remediation techniques, such as pulsed pumping and in situ bioremediation, enhance the efficiency of contaminant extraction and treatment, particularly for challenging organic pollutants in various subsurface environments. Nitrate contamination in shallow aquifers arises from industrial, agricultural, and municipal sources, necessitating integrated modeling of unsaturated and saturated zones to manage its transport. High arsenic levels in Western Anatolia highlight the need for advanced treatment methods and coordinated water resource management. Groundwater remediation strategies involve complex decisions between in situ treatments and extraction methods, requiring collaboration between hydrogeologists and engineers. Effective pollutant recovery techniques, such as aeration systems for VOC removal and understanding adsorption processes, are crucial for addressing contamination challenges in aquifers. Surface water and groundwater interact in complex ways, particularly within the hyporheic zone, influencing biological activity and nutrient cycling. Effective management of contaminated sediments requires quantitative modelling to link sediment concentrations to hazards, considering pollutants like heavy metals and PAHs that accumulate and bioaccumulate in aquatic ecosystems. Advanced sediment transport models must account for hydrodynamics, bioturbation, and uncertainty to accurately assess risks. Recent developments in groundwater monitoring and remediation design emphasize the integration of simulation and optimization models to enhance long-term management strategies despite inherent uncertainties.

PREFACE

More than 98% of freshwater on Earth is found in groundwater, which is also a significant component of terrestrial life. It releases nutrients and aids in the weathering of minerals in soils. Plants obtain their growth-promoting nutrients through the transpiration of soil water during photosynthesis. Even in some catchments, groundwater runoff into rivers sustains base flow vital to aquatic ecosystems. The human journey out of East Africa was facilitated by groundwater springs and oases, and ever since ancient times, towns have been constructed around dependable supplies of drinkable groundwater. The early hydrogeologists who constructed groundwater supply systems were concerned with the quality of the water as well as its source and quantity. We are getting close to the point at which we can no longer use this resource. In the recharge environments, the expansion of industries and cities poses a threat to potable groundwater. Over the past few decades, it has become clear that we need to conserve our groundwater supplies since they are finite. The development of geochemical and environmental isotope techniques to comprehend the source of groundwater, its solutes, and the mechanisms affecting groundwater quality has coincided with this. This book provides an introduction to groundwater concepts that can be useful to all students. Basic knowledge of water may help in understanding this book better.

INTRODUCTION

It is necessary to understand the processes of recharge and transport for groundwater development to be sustainable. This is necessary in order to comprehend the availability of groundwater and the relationship between groundwater and the environment. A conceptual grasp of groundwater flow at the local and regional temporal and spatial scales is necessary for management that will assist industry, agriculture, ecology, engineering, and all other environmental challenges. Initiatives for the sustainable development of groundwater require an understanding of the complete groundwater system. A local to regional understanding of groundwater flow could lead to this kind of awareness. Research on engineering, geography, agriculture, ecology, and, more broadly, environmental problems require an understanding of groundwater flow at the pertinent temporal and spatial scales. Research on the mobility of pesticides can generate insights for lowering leachate contamination on agricultural land.

Due to years of inefficient waste management and disposal practices at many sites worldwide, as well as the current lack of tools to adequately clean many of these sites, the remediation of contaminated underground systems is an expensive effort. We are unable to clean up contaminated subsurface environments due to a number of issues. Initially, areas with limited water permeability are frequently produced by subsurface heterogeneities. These low-permeability zones may become contaminated over extended periods of time when contaminants are released into the subsoil. A large number of organic pollutants are extremely resistant to chemical or biodegradation in the natural environment, and they are typically only able to be changed when they are in the aqueous phase. Many organic and inorganic contaminants frequently contaminate groundwater, and remediation techniques that are effective for one pollutant may not be suitable for another. It is rare for pump-and-treat remediation systems to be built for maximum efficiency. This book provides information about various issues in groundwater science.

CHAPTER 1

GROUNDWATER POLLUTION AND GEOCHEMISTRY

Contents

1.1. INTRODUCTION

The study of the mechanisms governing the chemical composition and quality of groundwater is known as groundwater geochemistry. The quality of the groundwater affects how this resource is used. Groundwater can contain unsafe materials that can harm people if they drink it or damage the environment if it is carelessly poured on the surface. The quality of the groundwater may alter as a result of exploitation or human activity, the effects of which are not always immediately visible. The primary goal of groundwater geochemistry is to guarantee clean drinking water. While it is possible to produce drinking water through desalination, it is still an expensive process. Environmental authorities/managers prioritize management of groundwater (Figure 1.1).

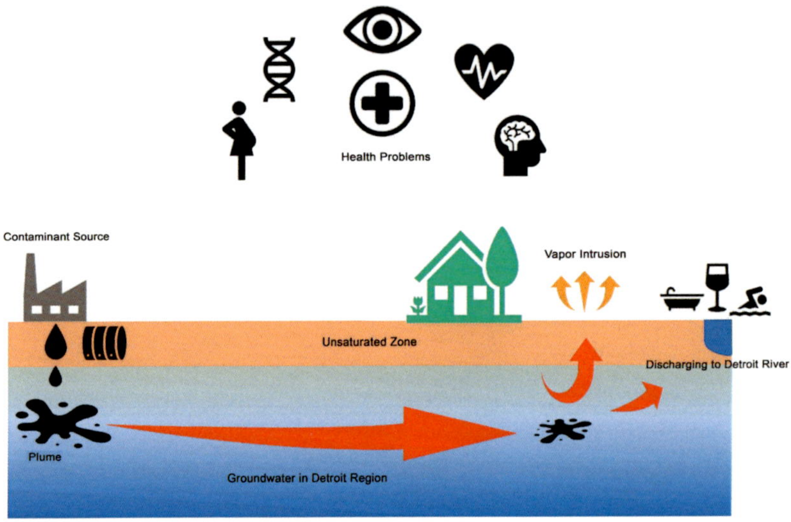

Figure 1.1. *Groundwater.*

Source: Teimoori, S., O'Leary, B. F., Miller, C.J. Modeling Shallow urban groundwater at regional and local scales: A Case Study in Detroit, MI. Water 2021, 13, 1515. https://doi.org/10.3390/w13111515, Creative Commons License.

In the upcoming decades, one of the most significant global issues will be the availability of clean freshwater. In 2003, one billion people lacked access to safe drinking water from public sources (Martínez-Santos, 2017). By the middle of this century, there will likely be between 2 and 7 billion people living in water-scarce countries (Biswas & Tortajada, 2019). According to recent predictions, almost 20% of the rise in global water scarcity would be attributed to climate change (Hanjra & Qureshi, 2010). Freshwater scarcity is a global issue that needs significant attention in order to guarantee that humans on the planet have access to clean drinking water. Most drinking water comes from surface water' however, groundwater is frequently chosen because it requires less treatment, which reduces the spread of water-borne diseases like cholera. This is common in underdeveloped nations.

1.2. GROUNDWATER SITUATION

The amount of Na we consume from drinking water is negligible compared to the amount of Na we obtain from the addition of salt (NaCl) to our diet. On the other hand, the intake of elements such as F and As through drinking water is very important, and an excessive concentration of these two elements poses a serious health risk to millions of people worldwide. Excessive fluoride consumption causes skeletal deformities known as fluorosis. This disease is prevalent in West Africa and India, where salts and sedimentary minerals containing fluoride are the main sources, as well as in African Rift Valley countries like Kenya and Ethiopia, where important volcanic sources of fluoride are. Approximately 67 million persons in India alone are susceptible to fluorosis (Kashyap et al., 2021). Inland basins with arid or semi-arid climates may contain groundwater with elevated arsenic levels. The aquifers are located in relatively flat, low-lying locations with little groundwater flow in some places. They are also situated in geologically young layers. Because arsenic is linked to sulfide minerals, mine waste water may have significant levels of arsenic. Skin lesions, skin cancer, liver disease, and hyperkeratosis can all be consequences of chronic arsenic poisoning. An estimated 20–70 million people in Bangladesh are at risk of consuming groundwater contaminated with high levels of arsenic (Raessler, 2018).

The application of water analysis in hydrology is to create a map that displays the regional distribution of the water components. These maps help practitioners such as drilling operators, water resource managers, environmental authorities/managers, and others locate aquifers with high-quality groundwater. They are also helpful for evaluating the relationship between the mineralogy of the aquifer and the composition of the groundwater. For instance, higher quantities of calcium and bicarbonate are probably present in groundwater drawn from a limestone aquifer. A typical groundwater study will include eleven variables: temperature, EC, pH, four major cations (Na, K, Mg, and Ca), and four significant anions (Cl, HCO_3, SO_4, and NO_3). Since rainwater is the primary source of groundwater, studying groundwater geochemistry makes sense. Oceanic vapor predominates in continental rains, which is similar to heavily diluted seawater. On the other hand, man-made and natural gases and dust change the composition. Different activities in the soil may alter the concentrations before the rain transforms into groundwater. Concentrated solutes are produced by evapotranspiration, and important components are chosen by vegetation and briefly stored in the biomass. The weathering of minerals in the soil plays an important role in varying concentrations, and the combined effects of soil activities frequently contribute to the chemistry of groundwater. Calculations of mass balance are helpful in gaining an understanding of the chemistry of water.

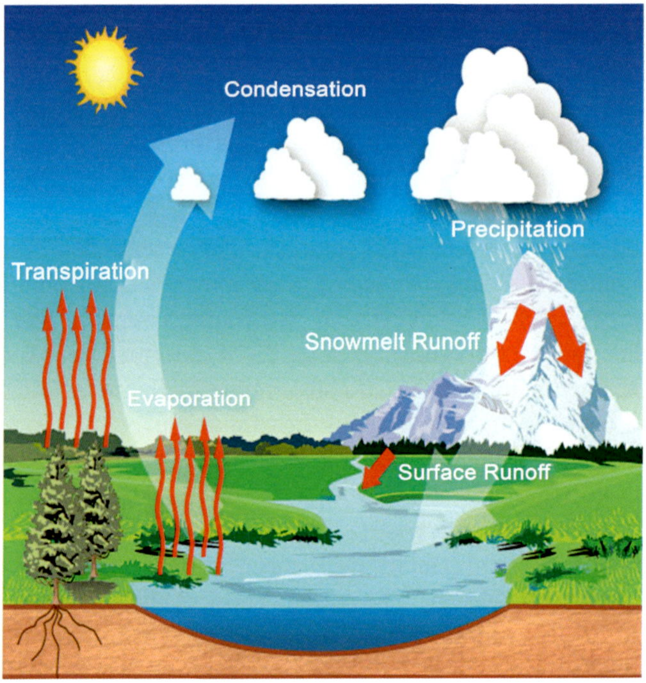

Figure 1.2. *Hydrological cycle.*

Source: https://www.noaa.gov/jetstream/atmosphere/hydro, Creative Commons License.

We can apply the fundamental and widely applicable mass balance approach to understand the hydrological cycle (Figure 1.2), the origins of the components in rainwater, the isotopic evolution of rainwater, the cycling of elements in the biosphere, and to determine the quality of groundwater beneath soils. Because local emissions disperse quickly and are transported throughout the world by the atmosphere, atmospheric pollution affects people on a global scale. Trajectory studies, which include tracking atmospheric particles, also known as air parcels, back in time using meteorological observations, have provided evidence of the transport velocity in the atmosphere. The findings indicate that it only takes roughly 4 days for particles or gases to move from the United States to Europe (Valavanidis et al., 2008). It is obvious that borders do not control the amount of industrial pollution in the atmosphere.

Both the adsorption of atmospheric gases and the deposition of aerosols are considered forms of dry deposition. Assessing the influence of dry deposition to the total atmospheric input of Cl is crucial because the Cl concentration in groundwater is being used as a parameter for estimating the recharge. Bulk samplers gather aerosols and anomalies, such bird droppings on the funnel that wash into the sampling container with the next rain, during dry spells. Volume samplers overestimate the input of the main ions by a factor of roughly 1.3 in the temperate Netherlands, where 800 mm of rain falls on a regular basis (Bouwman et al., 1997). When estimating the atmospheric input of various elements to the aquifer, the distinction between sampling types is also crucial because dry deposition is dependent on various factors such as wind direction, vegetation, and surface characteristics, all of which are difficult to extrapolate from the rain gauge to the natural surface. It displays the total anions, basically, chloride in the soil water. High rates of dry deposition result in high ion concentrations in the unsaturated zone. The air is filtered as one moves closer to the forest core, where the concentration drops by up to a factor of 10. Elements can be obtained from dry deposition and the majority of the aluminum used in dry deposition comes from soil dust, and its dry deposition flux varies depending on the region from 13 to 380 $mg/m^2/yr$ (Lisitzin, 2011). Trace elements are related to aluminum in the same proportion as those found in soil, and in many situations, human emissions have led to additional surges. Trace elements like Cd, Cu, Pb, and Zn are deposited dry across the North Sea at a rate roughly comparable to that of wet deposition. The overall atmospheric entry of Pb into the North Sea is roughly equivalent to the total riverine input, whereas the atmospheric input of Cd and Zn is only about 1/3 of the riverine input (Thevenot et al., 2007). Significant amounts of NO_x and SO_2 are also contributed by dry deposition' however, as gaseous adsorption plays a major role in the deposition of these components, dry deposition is often higher in coniferous canopies than in deciduous canopies.

Chloride concentrations are higher and can vary greatly below the forest, although they are typically lower below the rooting depth. The action of dry deposition helps to explain some of the high amounts of Cl under the forest. Furthermore, there are notable differences in evapotranspiration between wooded and grassland regions. Specifically, evaporation occurs more frequently in forests and can account for up to 40% of the precipitation. As a result, the concentration of dissolved ions increases. There was almost a three-fold increase in recharge rate beneath heathland compared to forest. The rain is only partially interacting with the more concentrated soil solution. The cations that are most prevalent are Ca^2, Na, and Mg^2 and HCO_3, SO_4, and Cl are the principal anions. Fe_2O_3, Mn_2O, and several trace metals are minor components. Anions like F and boron are also present in ground water. The distributions of several elements, like Na, SO_4^{2}-, and Cl, resemble the sigmoidal form of a normal distribution. This implies that the concentration of these species is determined by their availability in rocks, by the slow dissolution of minerals, or by biological processes. The slope steepens at increasing concentrations for other components, such as Ca^2, HCO_3, SiO_2, K, and F. This could mean that the maximum concentration of the species in natural waters is limited by the solubility of a particular mineral. Not only can organic matter oxidize in the soil, but it can also occur in aquifers that include fossil organic matter like lignite, peat, and other materials. Other electron acceptors, such as nitrate, iron oxides, or sulfate, may mediate the oxidation of organic matter once oxygen is depleted. Groundwater can contain substantial amounts of iron due to the reduction of iron oxides, sulfur in hydrogen sulfide (H_2S), which has an unpleasant odor, and methane in groundwater due to the fermentation of organic materials. The oxidation of organic materials, which produces CO_2, may also accelerate the weathering of silicate and carbonate minerals.

Figure 1.3. *Rock weathering.*

Source: https://en.wikipedia.org/wiki/Weathering#/media/File:KharazaArch.jpg, Creative Commons License.

1.3. WEATHERING

Rock weathering (Figure 1.3) releases chemicals into the water. The composition of the rain water in northern Italy as an example of the relationship between rock type and water composition. The rock composition exhibits a typical sequence of mica schists, limestones, dolomites, gypsum strata, and serpentinite rock found in the central alpine region. Alpine valleys incise the rocks, which dip abruptly. One can infer various rock weathering mechanisms using some tools. The typical catchment will have 10% water-filled soil that is one meter thick (Rodhe & Killingtveit, 1996). Water is thus present in the soil profile at 0.1 m. When the amount of precipitation exceeds evapotranspiration by 0.1 m/yr, the water will take a year to pass through the soil (Mu et al., 2022). The water travel time would be six months if the excess precipitation was 0.2 meters per year, etc., As a result, the annual runoff and the residence period of water in the soil are inversely correlated' the shorter the water-rock contact time, the higher the annual runoff. Additionally, minerals may precipitate in aquifers or soils. Secondary clay minerals are typically formed as a result of silicate mineral weathering. The bedrock composition and the degree of weathering determine the kind of clay mineral that develops. Al-silicates, either with or without additional cations, make up the majority of clay minerals. The clay minerals lose silicon and cations as weathering progresses, leaving behind only intermittently soluble Al-hydroxide. Fe-oxides are also expected to occur if iron is present in the parent rock. A variety of minerals may precipitate in the soil as a result of evaporation in arid conditions. These consist of chloride salts, gypsum, and calcite. Variations in solute concentration and minerals that precipitate during evaporation can be computed.

When the incoming water is different from the existing water, ion exchange processes between dissolved cations and those sorbed to mineral surfaces can significantly alter the water chemistry. Examples include pollution plumes extending from waste dumps, acid rain penetrating the soil, and fresh and salt water displacements in coastal aquifers. Ion exchange can result in concentration surges, but it also tends to smoothen concentration gradients caused by shifting conditions in the groundwater reservoir. Iron oxides, weathered primary minerals, clay minerals, and organic matter make up the sediment exchange complex in a soil.

Different types of water combine in coastal zones where fresh and saltwater mix, as well as in seepage zones for groundwater. Pollutant transport in aquifers also depends on the dispersion of various fluids. Even when the original fluids were at saturation before mixing, the process can nonetheless result in sub-saturation (calcite). The quality of water can be significantly altered by human activity. These contributions could be anything from domestic NaCl to heavy metals or toxic organic materials that seep from waste loads. In many countries, the levels of NO_3 and pesticides that seep from agricultural fields represent a major issue. Aquicludes, or aquitards when nearly impermeable, are layers of clay, peat, and loam that are thought to be impermeable. The aquifer is restricted if its water table is higher than the aquiclude limit' otherwise, it is unconfined and may not be saturated in its upper portion. An aquifer is said to be phreatic if it reaches the surface of the ground directly.

1.4. CONDUCTIVITY

If the hydraulic conductivity and gradient are known, Darcy's law can theoretically be used to compute the groundwater velocity (Sanchez-Vila et al., 2006). However, the computed velocity is somewhat unclear because of the wide variance in hydraulic conductivities. The water balance derived frequently offers a more accurate estimate of the useful parameters. Unlike total porosity, which includes the pores filled with stagnant water, effective porosity only contains the porosity containing water that participates in flow. Similar to the discharge and hydraulic gradient quotient, the effective hydraulic conductivity averages small changes in hydraulic conductivity. The water balance and the observed groundwater potential are utilized in numerical models to maximize transmissivity, which is the result of multiplying hydraulic conductivity by aquifer thickness. Transmissivity is a suitable metric if the amount of water flow is the sole factor. On the other hand, growing pollution has sparked curiosity about identifying the aquifer flow paths. Precise flowline assessment by numerical hydrological models has proven challenging due to permeability changes that dominate flow paths but have no effect on hydraulics. The maximum concentration of a mineral constituent parts in water is limited by its solubility. For instance, the mineral fluorite, whose composition is CaF_2, can be used to forecast the ranges of Ca_2 and F in groundwater concentrations by performing solubility calculations. Certain minerals react quickly with water. This is especially true for the more soluble minerals like fluorite, gypsum, halite (NaCl), and most carbonate minerals. Equilibrium will be reached in a shorter amount of time than groundwater residence period. Reaction kinetics must be taken into account because some minerals, usually silicates, react so slowly that equilibrium is never reached at low temperatures.

While low levels of fat in drinking water have been shown to be advantageous in some countries, and in fact, fat is added to water in many places where groundwater does not contain fluoride, high levels of fat pose a health risk, leading to dental fluorosis and skeletal fluorosis. There are high quantities in groundwater worldwide, but particularly in Asia and Africa. Basically, minerals in the rocks leach fluoride into the groundwater. Although the breakdown of shark teeth, which contain fluorapatite, has been identified as a source, volcanic rocks may also naturally contain significant levels of fluoride. Evaporation may raise the concentration of water in arid regions. Ion pairs and aqueous complexes can form when ions in an aqueous solution adhere to one another. Major cation complexes such as $CaSO_4$, CaF, or CaOH are examples. Heavy metal complexes such as CdCl, $HgCl_3$, or PbOH as well as metal complexes with an organic ligand such as oxalate, such as CuC_2O_4, are also present. The solubility of minerals and the mobility of trace metals will both increase as a result of complexation, which will decrease the activity of the free ion in water. The solubilization of mercury from the rock was caused by the complexing of Hg_2 and Cl, which raised the Hg concentration in groundwater in a coastal aquifer. Strong chelating agents like EDTA (ethylene-diamine-tetra-acetate) can help alleviate heavy metal poisoning, and complexes may have detoxifying properties. Complexation is used extensively in analytical chemistry, separation technologies, and the manufacturing of pharmaceuticals and food. Most complexes are made up of a metal cation that is surrounded by many ligands.

Figure 1.4. *Dolomites.*

Source: https://en.wikipedia.org/wiki/Dolomite_%28mineral%29#/media/File:Dolomite_Luzenac.jpg, Creative Commons License.

1.5. AQUIFERS

Dolomites (Figure 1.4) and limestones (Figure 1.5) frequently create productive aquifers that are ideal for drawing groundwater. The primary minerals found in these rocks are Ca- and Mg-carbonates, which readily dissolve in groundwater and contribute to the hardness of the water (Singh et al., 2016). Biological material that has been recrystallized and has a high porosity and low permeability typically makes up carbonate rocks. The more permeable fracture zones and karst channels created by the decomposition of carbonate can limit groundwater flow. In addition to carbonate rocks made entirely of carbonate minerals, carbonate minerals can also be found in sand, sandstones, and clays as accessory minerals or as cementing substances. The impact of carbonate dissolution on water compositions is noticeable all over the world. The crystal structure of carbonate minerals can be either orthorhombic or trigonal, contingent upon the cation ionic radius. Larger ions form orthorhombic minerals, while lower ions generate trigonal minerals.

Figure 1.5. *Limestone.*

Source: https://commons.wikimedia.org/wiki/File:Limestone_Eocene_deposit_at_Sinj_Stari_grad_-_Dalmatia_-_Croatia_IMG_20210820_083857.jpg, Creative Commons License.

When it comes to the trigonal mineral calcite, which is the primary constituent of limestones, the orthorhombic aragonite, which is frequently found in modern carbonate sediments, and the mineral that forms pearls, the Ca^2 ion is at the core or center. The carbonate minerals aragonite, dolomite, and calcite are those that form rocks. The other carbonate minerals, such as rhodochrosite or siderite, are rare and only found in trace amounts. Nevertheless, they can have a significant influence on the dissolved concentrations of Mn_2 and Fe_2. Different carbonate minerals are frequently found in solid solutions. Within the limits set by the degree of deformation permitted in the crystal lattice, solid solution can ascend or rise between minerals that are members of the same crystal group. Sr atoms can replace Ca in aragonite and Mg atoms can replace Ca in calcite' however, due to the extreme differences in ionic radii, Sr cannot replace much of the Ca in calcite. Mg-calcites are solid mixtures of magnesite and calcite.

Dolomite is another important carbonate mineral that forms rocks. The crystal structure of dolomite exhibits a great degree of organization, with the Ca and Mg atoms situated in distinct layers. This contrasts with the structure of Mg-calcite, where the Ca layer has atoms of Mg and Ca arranged randomly. Dolomite should be regarded as a pure state thermodynamically, whereas the laws of solid solutions apply to magnesium calcites. However, dolomite exhibits solid solution just like the majority of other minerals. The Ca/Mg ratio can change, and Fe and Mn substitution is frequent. The origin of dolomite is still debatable, however its highly compact structure makes it difficult to precipitate from aqueous solutions at low temperatures. Appropriate measurements of pH and alkalinity are necessary for a reliable estimation of CO_3. Field pH measurements can occasionally be unreliable due to several factors such as electrode temperature variations, buffer and sample variations, CO_2 degassing, malfunctioning or slow-moving electrodes, etc., The computation of CO_3 activity from alkalinity is directly impacted by errors in pH measurements. An alternate method is to measure the TIC and alkalinity. The pH and aqueous carbonate speciation can then be computed using PHREEQC (Schott et al., 2016). The recorded pH values tend to be slightly higher at low pH, where alkalinity pH buffering is least effective, possibly because of some degassing. For soils in temperate climates, the mean residence time of organic carbon is rather low, ranging from 10 to 100 years (Poeplau et al., 2021). The majority of the carbon dioxide generated in soil escapes into the atmosphere, where it remains for roughly ten years. The short residence period makes it more difficult to evaluate the global CO_2 budget and the impact of CO_2 on climate change. Changes in agricultural practices may have significant effects on the atmospheric CO_2 level, similar to the combustion of fossil carbon.

1.6. CARBON DIOXIDE

The concentration of CO_2 gas in unsaturated sandy soils under various vegetation should be studied. The forests have a greater CO_2 content and are made up of a mixture of hardwood and conifer trees. A steep gradient of the CO_2 distribution towards the surface (0 m depth) in soil profile denotes a CO_2 loss to the atmosphere. This CO_2 leakage was calculated to be 77 Gt C/yr on a global scale (Wen et al., 2024). The biological CO_2 generation stops in the winter, and the groundwater degasses to the atmosphere. The generation of CO_2 is highest in the summer, and the dispersion of CO_2 indicates both upward and downward diffusion. Because of the photosynthesis of terrestrial plants, the CO_2 concentrations in the atmosphere also vary seasonally. They are lowest in the late summer. The CO_2 content in soil air rises with temperature because of increased biological activity. But temperature cannot be the only factor affecting biological productivity' soil conditions and water availability must also be taken into consideration. Evapotranspiration (Figure 1.6) has thus been proposed as a substitution for biological production and, consequently, for the average CO_2 pressure in the soil.

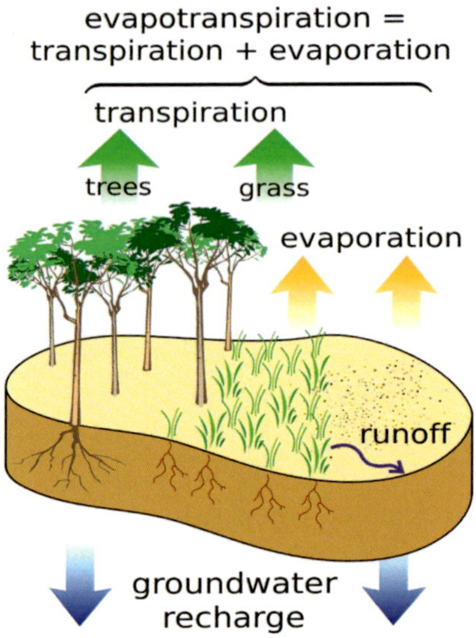

Figure 1.6. *Evapotranspiration.*

Source: https://commons.wikimedia.org/wiki/File:Surface_water_cycle.svg, Creative Commons License.

Dry soils cause an increase in CO_2 diffusion and a decrease in CO_2 concentration. Calcite precipitates as a result of the rising Ca^2 concentration brought on by gypsum dissolution' this process is comparable to the way gypsum dissolution affects fluorite precipitation. As calcite precipitates, the CO_3 concentration falls, which causes the dolomite to dissolve and the Mg^2 concentration to rise. Ca^2 must increase in tandem with an increase in Mg^2. Overall, the breakdown of gypsum causes the dolomite to change into calcite and results in fluids with higher amounts of Mg^2, Ca^2, and SO_4. Aquifers with dolostones, or carbonates of the dolomite type, connected to gypsiferous layers may experience dedolomitization.

The amount of gypsum melting can be tracked using the sulfate concentration. Regardless of the sulfate contents, the groundwater is either slightly supersaturated or at saturation for calcite throughout the aquifer. Gypsum can be in an equilibrium state or a substantially subsaturated state. Maintaining equilibrium with both dolomite and calcite requires that the [Mg^2] / [Ca^2] ratio remain at 0.8. It appears that surface saltwater is supersaturated in terms of dolomite, aragonite, and calcite. These minerals do not precipitate extensively from saltwater. Dolomite does not appear to precipitate at all in the oceanic surface layers, whereas calcite and aragonite do so very slowly. Seawater has almost five times as much magnesium as calcium. In a series of calcite and aragonite precipitation tests, researchers demonstrated that at concentrations of seawater, the rate of calcite precipitation in the presence of Mg^2 becomes negligible. Mg^2 apparently interferes with calcite crystal formation. A different theory is that adding Mg^2 increases the solubility, which prevents the formation of calcite. When supersaturation rises to high levels, the Mg^2-ion can be integrated, and the resultant calcites may contain as much as 30% $MgCO_3$. On the other hand, the addition of Mg^2 does not alter the crystal structure of aragonite. Aside from phosphates, organic acids are also well-known inhibitors that may prevent aragonite from precipitating. The Yucatan Peninsula in coastal aquifers were formed by the displacement of marine sediments into the freshwater zone due to sea level fluctuations caused by glaciers and deglaciations, which can vary by several hundred meters. When fresh water comes into contact with it, aragonite and high-magnesium calcite dissolve and a massive recrystallization process begins, causing calcite to precipitate.

1.7. ABSORPTION

Materials that can absorb chemicals include clay minerals, organic matter, and metal oxy-hydroxides found in soils and aquifers (Gomes & Silva, 2007). Adsorption describes a chemical's adhesion to a solid surface, absorption implies that the chemical is absorbed into the solid, and exchange is the substitution of one chemical for another at the solid surface. Differentiating between sorption and other solid-state events like dissolution and precipitation can be challenging at times. One significant distinction between sorption and other processes is that, unlike precipitation, sorption is dependent on the underlying solid surface. Hydrologists focus on sorption and ion exchange because they control the movement of harmful substances through soils and aquifers. A cation exchanger composition will be in balance with the local groundwater under steady-state conditions. The cation exchanger readjusts the composition of water to the new groundwater concentrations when the water composition changes due to pollution, acidification, or a changing salt/fresh water interface. By using a technique called ion-chromatography, the exchanger functions as a buffer that can drastically shift the concentrations in water. The hydrochemical patterns that arise from cation exchange can be used by a hydrologist to interpret hydrogeological conditions. The water quality along flow lines will vary as aquifers are used extensively and become more contaminated, but because of cation exchange, the water composition may still have certain characteristics of the previous water quality. A positive charge deficit in clay minerals can result from cation substitutions within the crystal structure. The length of the cation-oxygen bond plays an important role in filling the cation sites in the framework of oxygens. The length holds greater significance than the cation charge.

1.8. MINERALS

Mica-type clay minerals are primarily the only ones that exhibit a charge imbalance. Ion-chromatography is a common laboratory technique that uses the concepts of ion exchange to separate and examine ions from complex mixtures. The chemical analyst employs an ion-exchanger column filled with beads. Fluid fills the pores in and between the beads' the total volume of fluid in the column is known as the column pore volume, or pore volume.

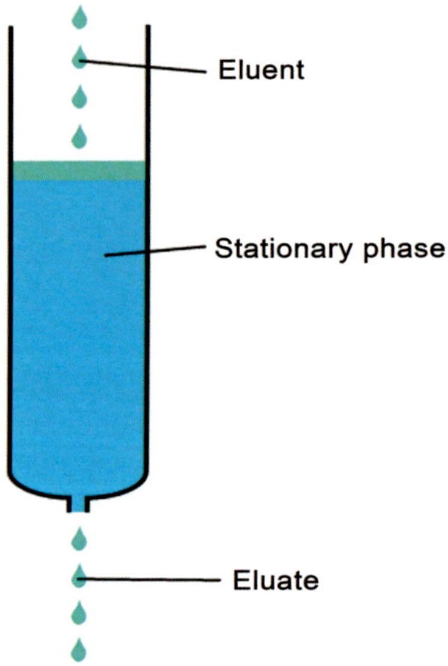

Figure 1.7. *Elution chromatography.*

Source: https://commons.wikimedia.org/wiki/File:Chromatography.png, Creative Commons License.

A concentration gradient forms at the end of the column as a result of cation exchange after the column is percolated with various solutions. Ions will move other ions out of the exchanger and move at a slower speed. The cation in the injected solution is sorbed in chromatography than the resident cations. The least selective ion is leached first, followed by the next most preferred, and so on, until the cation that is injected into the column eventually gets spotted. Elution chromatography (Figure 1.7) is the procedure used if, on the other hand, the cation in the incoming solution has a lower affinity for the exchanger. In a porous media, the flow conditions are frequently non-uniform, with water circulating in specific areas and remaining stagnant among particle aggregates. There will be a diffusive ion exchange between the mobile flow region and the stagnant zones. Particle aggregates may experience intraparticle diffusion, a somewhat slower process. The size of the stationary zone in a homogeneous sand correlate to the pore size, which is roughly equal to the particle size below which 10% of the particles reside. Sorption processes play a major role in controlling the mobility of trace metals after they are discharged into groundwater.

1.9. SORPTION

Sorption, however, is associated with the specific binding of heavy metals to organic matter and oxides that have varying charge surfaces (Naidu et al., 1998). Ions are absorbed from solution by variably charged materials without releasing an equal amount of other ions. Depending on the pH and composition of the solution, their surface charge might be either positive or negative. The mobility of positively charged heavy metals like Pb_2 and Cd_2 is controlled by variable charged solids. Compared to ion exchange, modeling sorption on variable charged solids is more complicated and typically calls for computer programs to provide an accurate result. Essentially, protons and other ions from solution sorption give the surfaces of oxides and hydroxides a pH-dependent charge. A potential difference between the surface and the solution is produced by the surface charge. This potential relates back to the surface charge increase via influencing the ions approaching the surface. Nonetheless, certain ions have strong bonds that they even stay attached to a surface that repels the other ion electrostatically and has the same charge. Large volumes of heavy metals that are still soluble in the acid mine drainage are simultaneously released. The water in the upper portion has a pH of less than 2.5 and more than 20,000 milligrams of zinc and 100 mg of copper per liter (Cravotta, 2008). Zn stays high and only lowers in the saturated portion, while the pH rises to over 4 and the Cu content falls to 1 mg/L as one moves downward. The mobility of the heavy metals appears to be influenced by pH, albeit the effect varies depending on the metal. Acid mine drainage is an environmental problem. Many abandoned metal sulfide mines regularly discharge low pH and heavy metal-laden water into the environment. Pumping down the water table may allow sulfides that were previously submerged and shielded from oxidation to access ambient oxygen. A high sulfate concentration is present along with a high nickel concentration, indicating that pyrite oxidation is likely occurring. Since pumping began in some places, nickel has not advanced much in the aquifer, presumably because sorption on iron oxyhydroxide and manganese oxides slows its flow.

1.10. METALS

Trace metals may also be released by the chemical weathering of oxyhydroxides, clay minerals, and detrital silicates like biotite and amphiboles. Atmospheric input may be significant. The heavy metals are fixed by sorption or solid equilibria at neutral pH, but they can be mobilized when water becomes acidic. Acid rain is making the upper groundwater pH of 4.5 and elevated levels of heavy metals like Ni, Co, and Be (Rupias et al., 2021). The pH rises to about 6.5 further down, and the amounts of heavy metals are lower. With half-lives of only a few minutes, trace metal sorption on humic acid or oxides occurs rapidly. Long reaction times, however, have the potential to improve sorption. Desorption requires more time than adsorption. By tracking the time-dependent variation in the species concentrations, we can attempt to simulate a kinetics experiment and determine the various processes. The weathering of silicate minerals is thought to be responsible for roughly 45% of the total dissolved load (rivers). One important CO_2 sink in the global CO_2 cycle is the weathering of silicate minerals. Moreover, the primary method of pH buffering in sediments devoid of carbonate particles is silicate weathering. Aquifers in silicate rock are susceptible to acidity due to the slow pace of silicate breakdown. Moreover, silicate weathering as a pH buffer has drawn attention since acid rain is causing forests to deteriorate.

Soil weathering of detrital silicate minerals has been investigated. The change in mineralogical composition that arises as a function of depth and time shows even very slow processes of mineral degradation and transformation. Soils may be subjected to chemical weathering for thousands of years. Plagioclase makes up 32% of the parent granodiorite rock, followed by quartz (28%), K-feldspar (21%), biotite (13%), muscovite (7%), and trace levels of hornblende (2%) (Basta et al., 2017). While they appear to be less resistant to weathering in saprolite (partially weathered rock), biotite and K-feldspar appear to be more resistant. The progressive loss of distinct silicate minerals during weathering is indicative of the overall kinetic control over the distribution of primary silicate minerals. The minerals that weather the easiest are olivine and calcite, while the minerals that weather the hardest are quartz and ilmenite. When it comes to weathering conditions at the surface, the minerals that developed at the highest temperature are the most unstable. Some secondary minerals are insoluble byproducts that arise from the breakdown of primary silicate minerals in an inconsistent pattern. Dissolution occurs when the ratios of the elements that appear in the solution and the dissolving mineral are different. The addition of cations and silica is the main impact of silicate weathering on the chemistry of the water. Almost all silicate weathering processes raise pH levels by consuming acid. Organic acids and carbonic acid are the main sources of protons in an unpolluted environment. Silicate weathering will result in the production of bicarbonate. Iron found in silicate minerals such as hornblende or biotite may precipitate Fe-oxide, an insoluble weathering product.

1.11. SILICATE

Silicate rocks and sandstones that have been percolated by groundwater will also undergo weathering reactions; however, the mineralogical changes in these materials are frequently harder to spot than in soils. The high silica content suggests that silicate minerals are actively degrading. Rhyolite and basalt, two volcanic rock types that contain more reactive material than mica-schists or granite, have the highest concentration. Sodium plays a major role in the cations in all groundwaters, and its presence is not counterbalanced by chloride, which is what would be expected if Na came from seawater. Albite and other members of the plagioclase solid solution series between albite and anorthite (Ca-feldspar) are the primary sources of sodium through weathering. Both plagioclase and hornblende, also known as amphiboles, and pyroxenes weather, releasing $Ca^{2+.}$ An increase in cation concentration is correlated with an increase in bicarbonate, and it seems that silicate weathering may occasionally even result in carbonate precipitation. Hawaii's soils, where bauxite, kaolinite, and montmorillonite form at varying concentrations with altitude on the volcanic slopes, provide a typical illustration of how precipitation and leaching intensity affect the composition of weathering residues. Montmorillonite (Figure 1.8) is found in regions with minimal rainfall because the water has a longer residence time in the soil, increasing the solute concentrations. On the other hand, bauxite occurs in regions with heavy rainfall, high leaching, and short soil residence times for water, resulting in low concentrations of dissolved ions. Since the kind of clay mineral can have a significant impact on the permeability of the rock, sedimentary petrology has examined clay mineral formation resulting from the alteration of primary silicates in great detail. It has been demonstrated that sandstones cemented with illite have permeabilities at least orders of magnitude lower than sandstones cemented with kaolinite, although having the same porosity.

During erosion and transportation, clay minerals that are the result of silicate weathering stay relatively stable and are deposited as coatings on sand grains, frequently in close proximity to iron oxides. These coatings have a significant impact on the aquifer ion exchange characteristics. Occasionally, clay particles are carried further into the soil profile by runoff from the higher soil layers. The overall percentage of clay found in sand deposits is often less, and it is typically made up of a variety of different clay minerals. Reactions between soil and aquifer components may neutralize acid water put into soil and aquifer systems.

The relationship between the geology and the susceptibility to acidification must be taken into account in order to forecast the amount of neutralization. Because carbonates dissolve quickly, it follows that rocks containing carbonate minerals are unlikely to produce acid groundwater. A gradual decline in alkalinity serves as a strong indicator of impending groundwater acidification. Granites, acid igneous rock, and clean quartz sandstones, which are primarily composed of slowly dissolving silicate minerals, were listed as the carbonate-free rocks most susceptible to acidification, while basic and ultrabasic rocks, which are primarily composed of quickly dissolving silicates, were found at the other end of the scale. When the acid load is less than 0.5 keq/ha/yr, researchers state that the presence of traces of pyroxene, hornblende, or biotite should be sufficient to prevent acidification. Below a coniferous forest, a low reactive sandy aquifer exhibits groundwater acidification.

Figure 1.8. *Montmorillonite.*

Source: https://commons.wikimedia.org/wiki/File:Montmorillonite-Quartz-pala48a.jpg, Creative Commons License.

The natural quantities of O_2, Fe_2, SO_4, H_2S, CH_4, and other elements in groundwater are mostly regulated by reduction and oxidation processes (Christensen et al., 2000). Additionally, they ascertain what happens to pollutants such as heavy metals in acid mine drainage, pollutants from industrial spills, nitrate leaking from agricultural fields, and toxins leaching from landfills. Redox reactions are caused by the transfer of electrons from one atom to another, and thermodynamics can be used to predict the sequence in which these events will occur. However, the electron transfer is frequently extremely slow and may only occur when bacteria catalyze the process. The reduction of sulfate by organic matter, which occurs in marine sediments and aquifers, is one example. Abiotically, the reaction is incredibly slow' but, in natural situations, the reaction occurs quickly because bacteria such as *Desulfovibrio* sp. produce the enzymes needed to catalyze the activity. Groundwater undergoes redox reactions primarily when an oxidant (O_2 or NO_3) is added to an aquifer that already contains a reductant.

But it can also be beneficial to add a reductant, like dissolved organic matter, which leaches from landfills or soils. Downstream from landfills, redox zoning may occur in the reverse direction. The various redox zones are distinguished according to the composition of the water. The conditions are methanogenic' however, they alter downstream to reduce sulfate, iron, and manganese, then to reduce nitrate. The clean groundwater is oxic around 400 meters downstream of the landfill. The landfill or leachate plume gets less reduced downstream than it would naturally when precipitation infiltrates, and electron acceptors are needed to oxidize the organic materials in the plume. Only mixing, dispersion, and diffusion can bring dissolved electron acceptors like sulfate, nitrate, and oxygen into the organic-rich leachate. The organic leachate is combined with sulfate-enriched groundwater that is released by nearby sulfate-rich sources inside the landfill, such as gypsum waste. Similarly, only the mixing of nitrate-bearing groundwater with organic leachate can result in a nitrate-lowering zone. Carbonates may dissolve when a significant amount of organic material in the leachate oxidizes and produces a lot of CO_2. Due to infiltration of urban wastewaters, similar plumes of organic pollutants are frequently detected in aquifers under cities in developing nations.

1.12. NITRATE

Groundwater nitrate contamination is becoming an issue in all of Europe and North America. It is a serious threat to groundwater-based drinking water supplies. The recommended amount is less than 25 mg/L NO_3, while the permissible nitrate content in drinking water is 50 mg NO_3/L (Zhang et al., 1996). Due to its propensity to cause methemoglobinemia in babies, a potentially deadly disease characterized by impeded oxygen transport in the bloodstream, excessive nitrate concentrations in drinking water are thought to pose a health risk. Many point and nonpoint sources, such as septic tanks, cattle feed, sewage discharge, and the oxidation of organically bound nitrogen in soils, contribute to the groundwater nitrate level. Nonetheless, since the early 1960s, the overuse of fertilizers and manure in agriculture has unquestionably been the primary factor contributing to the rising nitrate content in groundwater. In natural situations, intermediates like

NO_2 and N_2O are frequently detected at trace quantities, which is taken as proof that denitrification is still occurring. N_2, however, is consistently the main reaction product. There are no microorganisms that can survive on the energy released by oxidizing N_2 to NO_3 since denitrification is an irreversible reaction. In groundwater systems, nitrate reduction to NH_4 can occur. The process of nitrification, which is significant in the soil zone, involves bacteria oxidizing amines from organic materials to nitrite and nitrate. The primary element influencing the overall rate is most likely the reactivity of organic materials.

In the oxidized zone, the O_2 level is almost constant and almost at air saturation. Oxygen and nitrate both disappear simultaneously at the redoxcline, therefore the reduction process needs to occur quickly in comparison to the transport rate, which is roughly 0.75 m/yr. According to the distribution of pyrite in the sediment, increases in sulfate and Fe_2 reflect the oxidation of pyrite. Iron is present in the traces as iron oxide coatings on the grains of quartz and feldspar. Primary Fe^2-bearing silicates, such as biotite, pyroxenes, and amphiboles, are present. Such minerals often dissolve at very low rates, which means that the resulting Fe_2 concentration will be low as well. Amphiboles and pyroxenes were discovered in a sandy aquifer with a Fe_2 concentration of 10^{-13}M and unique dissolution characteristics (Canfield et al., 2005). The iron oxyhydroxide layer that forms from the Fe^2 produced during dissolution under oxic conditions prevents further dissolution. Ferrihydrite, goethite, lepidocrocite, and hematite are common iron oxides and oxyhydroxides found in sediments. When Fe^2-rich groundwater interacts with ambient oxygen, iron oxyhydroxides precipitate and produce deposits in river beds and iron ores. X-ray can be used to identify the iron speciation in aquifer sediments. Nonetheless, wet chemical extraction techniques are typically used to identify the iron speciation in sediments. The dissolving rate is significantly increased by organic molecules that chelate with iron at the oxide surface.

CHAPTER 2

GROUNDWATER MANAGEMENT

Contents

2.1. INTRODUCTION

When it comes to providing water, groundwater is preferred to surface water in a number of ways. It is typically of higher quality, more resilient to microbial and chemical pollutants, less prone to seasonal fluctuations, and more evenly distributed across large areas. In the United States, groundwater provides 33%, 98%, and 42% of the water used for irrigation, household use, and public supply, respectively (Fram & Belitz, 2011). In 2005, irrigation accounted for over 67% of fresh groundwater withdrawals, with the remaining 18% going toward municipal supplies (Brown et al., 2013). While groundwater use surpasses 70% of total water use and is the primary source of drinking water supply in most European countries, groundwater is the only source of water for several countries outside the United States. Groundwater plays an important role in maintaining vital ecosystems, riparian and other types of vegetation, and important surface water systems, even though, like rivers, its value has not been widely acknowledged. According to estimates, groundwater contributes roughly 36% of river runoff globally' however, in the United States, groundwater contributes even more to streamflow, with estimates of 52%. One of the most valuable natural resources is groundwater. It should come as no surprise that groundwater has been overused both domestically and internationally given the demands of the populace. Overutilization can have a variety of negative effects, such as declining water levels and aquifer depletion from groundwater pumping, surface water resource depletion from groundwater-surface interactions, land subsidence from groundwater development, and saltwater intrusion into coastal aquifers. Furthermore, poor waste management techniques have contaminated shallow aquifers with chemicals and microorganisms, lowering water quality and making the supply unfit for residential and commercial usage. Similar to this, regionally applied fertilizer, pesticide, and herbicide treatments have had a negative effect on shallow aquifer water quality. The state of our economy, the environment, and our quality of life all depend on the efficient management of resources, both in terms of quantity and quality.

2.2. MODELS

Models can be used to forecast how the system will react to any operation strategy and to determine the new state. In rich countries, the policy might take the shape of enforceable laws, whereas in underdeveloped nations, it might take the form of social or economic incentives designed to change the population behavior. Decision analysis can be used in both situations to determine the best locations, rates, and times for pumping or recharging in order to meet management goals and policy. Since improper disposal of waste has resulted in a decline in the quality of water in many aquifers, groundwater restoration has grown in importance as a part of groundwater management. Remedial technologies have been developed that can lower pollution concentrations to levels that would let the aquifer be used as a source of water supply, depending on the type of contamination. These remediation techniques can have a physical, chemical, or biological basis, but for their application to be successful, they must be compatible with the pollutants on the site as well as the subsurface features. According to projections, over half of the freshwater resources would be strained or in short supply by 2025. Trends in water shortage can be linked to both population expansion and climate change, according to models assessing how vulnerable water supplies are to future changes in the climate, migration, and industrial development. The blend of climate change and population growth to approximately 8 billion people is anticipated to cause an increase in water scarcity.

Because groundwater is essential to the water supply, future demand growth will pose serious problems for groundwater management. In fact, as groundwater replaces surface water for all purposes, domestic, agricultural, and industrial, the demand on the groundwater resource is predicted to be considerably higher than it is now. Advantages in terms of quantity and quality, along with its simple accessibility due to more affordable and efficient drilling and pumping methods, are what are causing this shift. Groundwater withdrawals are expected to remain relatively stable in the United States and Western Europe, but they are expected to increase most quickly in developing nations like China, Bangladesh, and India. Global trends also indicate that it is critical to concentrate on groundwater use in agriculture for the simple reason that quantities used in this sector vastly outweigh those used in industry and residential applications globally. According to FAO projections from 2003, when it comes to the use of groundwater for agriculture, access to it will continue to enable intensification of agricultural production in response to shifting demand patterns (Pfister et al., 2011). There is little room for managing the need for groundwater in agriculture, especially in rural areas where people are attempting to escape poverty. Overuse of aquifers by agriculture will force users into social and economic transitions. Although demand management, which strictly restricts the development and use of groundwater, is known to be difficult to implement in developing nations, it should be acknowledged that intensive groundwater exploitation in agriculture is an extensive global phenomenon that occurs in a variety of socioeconomic contexts and has similar effects on the environment. Survival and profit, the two main human motivations for groundwater development for irrigation, can be seen as extremes that are largely determined by the stage of regional development. The management of water resources is an increasingly difficult task due to the growing demand for water brought on by population increase and fast urbanization. Preventing and controlling groundwater pollution is one of the most crucial facets of managing water resources. Because groundwater quality and supply are crucial to sustainable development, groundwater contamination is a serious problem.

2.3. CONTAMINATION

Contaminants in groundwater are generally linked to detrimental effects on social, economic, environmental, and health aspects. Certain pollutants are linked to cancer, damage to the liver and damage to the central nervous system, which raises health concerns. There is a great deal of uncertainty regarding the health effects of exposure to many of the pollutants found in groundwater. Because polluted water can be tasteless, odorless, and colorless, exposure to contaminated groundwater can occur by accident. Degradation of water quality as a result of soil, surface water, and air interactions with contaminated groundwater (Figure 2.1) is one of the environmental consequences.

Figure 2.1. *Contaminated groundwater.*

Source: https://en.wikipedia.org/wiki/File:Pit_latrine_and_well_in_close_proximity_(3794846044).jpg, Creative Commons License.

For example, baseflow allows pollutants from groundwater to reach bodies of surface water. There may be negative consequences from this association for wildlife, fish habitat, and vegetation. The substantial expenses associated with identifying, stopping, and managing pollution are among the negative economic effects of groundwater contamination. In addition, the expenses associated with creating auxiliary water sources, potential mortality or morbidity costs, and a decline in industrial and agricultural output due to the use of the contaminated water can all be considered forms of economic damage. The fact that groundwater poisoning occurs below the surface is one of the main issues. Since the detection of contaminants typically relies on acquiring samples of groundwater from monitoring wells, which sample a very tiny amount of the aquifer, contaminants are difficult to observe. Because contaminated groundwater is typically colorless and odorless, detection may be challenging. Moreover, many pollutants have very low concentrations at which they represent a health risk, making analytical methods for detecting and quantifying them challenging. Long after the original incident that triggered the pollution, groundwater contamination still has significant effects.

The following conditions may determine the degree of groundwater contamination. The use of contaminated groundwater as a source of potable water' the concentration of contaminants in relation to the maximum level of contaminants' the number of people impacted by the contamination' the area contaminated by the contaminants' and the degree of hazard posed by the contaminants, which is dependent on the toxicity, persistence, concentration, and movement of the contaminants in the aquifer. For the purpose of accurately determining the level of groundwater pollution, evaluating the associated risks, and designing efficient remediation and protective measures, it is imperative to comprehend the destination and transit processes of groundwater contaminants. It is usually necessary to create and use mathematical models in order to characterize and control groundwater contamination. Models can be applied in determining the extent of groundwater contamination' evaluating the impact of current practices on groundwater quality' projecting future conditions under current practices' and assessing the efficacy of likely corrective measures, protection alternatives, and management strategies meant to minimize groundwater contamination.

2.4. SOURCES

Point and nonpoint sources are the two categories into which groundwater pollution sources can be divided (Chowdary et al., 2005). Leaking septic tanks, wastewater lagoons, underground pipelines, tanks, and barrels containing chemical wastes are examples of point sources. These sources may be identified, are confined to certain areas, and affect groundwater quality locally. On the other hand, non-point sources are dispersed, have a regional impact on groundwater quality, lack a clearly defined single point of origin, and originate from many smaller sources. Agricultural practices including applying fertilizer and manure, precipitation, and runoff from parking lots and roadways are a few examples of non-point sources. Reconstructing the loading history of groundwater contamination, both geographically and temporally, from the current spatial distribution is a standard approach in groundwater engineering. Groundwater can potentially be contaminated by nearly any kind of material, coming from a wide range of sources. The majority of pollutants are the result of improper garbage disposal into or onto the ground. Unintentional product leaks from underground and aboveground storage facilities could potentially be sources of contamination. There are four primary processes that lead to groundwater contamination: infiltration, direct migration, inter-aquifer exchange, and recharging from surface water. The sources of groundwater contamination were divided into four groups: industrial, municipal, agricultural, and other sources. Groundwater contamination sources were categorized into six groups (Ganiyu et al., 2021). Sources classified as

Figure 2.2. *Septic tanks.*

Source: https://en.wikipedia.org/wiki/Septic_tank#/media/File:Septic_tank.jpg, Creative Commons License.

Category I comprise septic tanks (Figure 2.2), cesspools, and land application of wastewater through sludges or spray irrigation. These sources are intended to release pollutants. Using the soil's ability to absorb wastewater is the fundamental idea behind the construction of these systems. Groundwater has been known to receive ammonium-rich residential wastewater discharged from septic tanks. Injection wells for wastewater represent an important path for contamination. Essentially, the purpose of injection wells is to release liquid wastes into underground areas that are underneath any aquifers that supply drinking water. Injection wells have the potential to contaminate groundwater due to poor well design and construction, potential pollutant migration to usable aquifers, and likely constraining bed faults and fractures.

Figure 2.3. *Landfills.*

Source: https://commons.wikimedia.org/wiki/File:Wysypisko.jpg, Creative Commons License.

Sources classified as Category II are those intended for the storage, treatment, or disposal of materials' they are not intended for the subsurface release of pollutants. Open dumps, landfills (Figure 2.3), residential disposal, waste piles, material stockpiles, animal burial grounds, above- and below-ground storage tanks, containers, open burning and detonation sites, and radioactive waste disposal sites are some examples of these sources. Sources intended to hold chemicals during transmission or transportation are under Category III. The main components of these sources are material transport and transfer operations and pipelines. Gas, petroleum, and wastewater can all be transported via pipelines. Sanitary sewer pipeline leaks have the potential to contaminate groundwater with bacteria, viruses, nitrate, and chloride. Sources that release compounds as a result of other scheduled activities are under Category IV. The majority of the activities in this category is related to agriculture and contain fertilizer and pesticide applications, and animal feeding operations. This category also includes other sources, like drainage from surface and underground mines, urban runoff, and the seepage of air pollution. Sources that arise from the unintentional formation of underground conduits and modifications to groundwater flow patterns fall under Category V. Installations of water supply wells, geothermal and heat recovery wells, oil and gas wells, monitoring and exploration wells, and construction dewatering wells can all produce these sources. Generally, these wells can act as vertical conduits for pollutants to enter groundwater from the surface or from underlying contaminated strata when they are built of inappropriate materials, incorrectly abandoned, or improperly completed. Additionally, conductive fissures that allow surface contaminants to enter might be exposed by building excavations that remove soil overload from bedrock. Naturally occurring sources that may result in groundwater contamination due to human activities are included in Category VI. Sources in this category include interactions between surface and groundwater, natural leaching, intrusion of saltwater, and brackish water (Figure 2.4).

Figure 2.4. *Brackish water.*

Source: https://en.m.wikipedia.org/wiki/File:Mangroves_Around_the_Brackish_Waters_of_the_G%C3%A1lapagos.jpg, Creative Commons License.

Aquifers are replenished from surface water when the stream base is higher than the water table in the groundwater-surface water interaction (Safeeq & Fares, 2016). Contamination from the surface water may seep into the aquifer upon recharge. Excessive pumping beyond the sustainable yield of the aquifer in coastal places causes salt water to seep into freshwater aquifers. Groundwater contamination and the leaching of natural components from the soil are two possible outcomes of irrigation. Radionuclides, trace elements, nutrients, other inorganic species, organic pollutants, microbiological contaminants, and other species make up the majority of contaminants. As long as these pollutants are present at concentrations higher than the maximum contamination limit (MCL), they all have the potential to seriously endanger human health.

2.5. RADIOACTIVE MATERIALS

Most groundwater contains naturally occurring radioactive materials. Depending on the underlying bedrock origins, the radionuclides that are present and the activities that go along with them differ. An additional factor in the radioactive inventory in groundwater is human activity. Tritium and fission products have been found in groundwater as a result of fallout from above ground nuclear weapons testing. Mine drainage/leak from uranium ore mining may contaminate groundwater. Radioactive waste is produced during the refining and recycling of fuel and might accidentally be emitted as solid and liquid waste. Groundwater pollution has been a result of effluents from nuclear power plants, which generate electricity by using radioactive materials. Groundwater contamination may arise from the storage of spent fuel and the disposal of other radioactive waste. It is widely known that exposure to radiation and radioactive pollutants pose health concerns. Some researchers list cataracts, non-malignant skin damage, bone marrow depletion, and infertility as other health issues. Aluminum, antimony, copper, gold, iron, lead, lithium, manganese, mercury, molybdenum, silver, strontium, thallium, tin, titanium, uranium, vanadium, and zinc are among the major contaminants known as trace metals. Because trace metals tend to accumulate in the body, they can be poisonous and hazardous to humans even at low amounts. The primary sources of trace metals in groundwater are runoff and solid wastes' wastewater from industrial and urban areas' agricultural activities, such as the application of fertilizer and manure' and fossil fuels. Nitrate (NO_3) and ammonium (NH_4) are the most prevalent nitrogen species in groundwater. Nitrate pollution of groundwater is regarded as one of the most serious contamination problems globally. Recent research demonstrates the ongoing presence of nitrogen in both soils and groundwater, despite decades of substantial study on the problem of nitrate contamination of groundwater and the development of many protective alternatives. Drinking water containing high levels of nitrate can cause methemoglobinemia in newborns and stomach cancer in adults. If nitrate comes from animal feces or septic system effluent, it could be a sign that bacteria, viruses, or protozoa are present in groundwater. Agricultural practices are also likely the main cause of nitrate contamination in groundwater (Bijay-Singh & Craswell, 2021).

2.6. AGRICULTURE

Fertilizers, dairy farms, manure application, and leguminous crops are examples of nonpoint agricultural techniques that provide supplies of nitrogen. The largest non-point source of nitrate leaching to groundwater, especially on sandy soils, is thought to be the heavy use of fertilizers on crops. When farms are over-irrigated, nitrates from the fertilizer seep into the groundwater in urban areas. When it comes to animal manure application, there may be a greater risk of nitrate movement to groundwater than with crops, particularly in regions with a large concentration of dairy farms. Groundwater around farmyards and dairy companies frequently has elevated nitrate concentrations. It has also been demonstrated that nitrate concentration rises when poultry manure is added. Non-point sources of nitrogen include dry deposition, irrigation flows, and dissolved nitrogen in precipitation in addition to agricultural operations. It has been demonstrated that groundwater nitrate pollution is a result of point sources of nitrogen. Many investigations have revealed significant nitrate concentrations near septic tanks. The problem of groundwater pollution by phosphorus is regarded as less significant than that of nitrate. This is because phosphorus tends to adsorb onto solids, which makes phosphorus molecules in groundwater less mobile and soluble. This category of pollutants includes nonmetals, which include ions containing carbon and sulfur or other species like fluoride and chloride, as well as metals that are present in non-trace amounts like calcium, magnesium, and sodium. A large number of these ions are primary causes of groundwater salinity. The water is unsuitable for many industrial uses and human consumption due to high quantities of these ions. High quantities of ions, like sodium, can abruptly alter the chemistry of blood and cells.

2.7. ORGANIC COMPOUNDS

The main component of organic molecules is carbon, while other common constituents include hydrogen and oxygen. Naturally occurring groundwater always contains dissolved organic material. Organic substances in groundwater can typically dissolve in water or exist as an immiscible liquid. Because of human activity, organic molecules are a major source of concern. Globally, a lot of organic compounds are used, particularly crude petroleum products. Soluble aromatic hydrocarbons are an important class of organic chemicals that seep into groundwater after petroleum fuel or lubricant spills (Kuppusamy et al.,

2020). Toluene, benzene, ethylbenzene, para-, meta-, and ortho-xylenes are all members of this category. This class of organic compounds is known as the BTX (benzene, toluene, and xylene) or BTEX (benzene, toluene, ethylbenzene, and xylene). Another important class of synthetic compounds produced by humans that frequently contaminate groundwater is represented by chlorinated solvents. These are usually aliphatic compounds that have undergone halogenation and have a mild solubility in water. The two most common types are trichloroethene (TCE) and tetrachloroethene, which is also referred to as perchloroethylene (PCE). Due to old disposal methods and unintentional discharges, these solvents were widely used in industry and have frequently contaminated groundwater. Chlorinated solvents are relatively immiscible and denser than water. When released into the groundwater, they accumulate on clay or bedrock surfaces and they are unable to penetrate and sink as a distinct liquid phase through the groundwater due to gravity, leaving trace amounts in the pore space as they descend. This distinct phase is referred to as DNAPL, or Dense, non-aqueous phase liquid. Because of its slow rate of breakdown, residual DNAPL can act as a long-term source of groundwater contamination. In general, drinking water contaminated with organic substances increases the risk of developing cancer, liver damage, heart problems, nervous system malfunction, brain abnormalities, and other lesions.

2.8. EPA

The US EPA was empowered to establish drinking water standards in 1974 and revised in 1986 and 1996. The US EPA establishes drinking water standards, which are rules that are applicable to all public water systems and aim to limit the amount of pollutants in drinking water. The Maximum Contaminant Levels (MCLs), which are enforceable health-based norms, are the primary standards. Secondary standards are non-binding guidelines that address aspects of the water esthetic quality. The Maximum Contaminant Level Goal (MCLG), as determined by the US EPA, is approximated by the MCL. The Maximum Contamination Level of a pollutant in drinking water is the highest level at which there are no known harmful effects on human health. The MCLG for carcinogenic pollutants is zero, which means that no level of contamination is permitted. The Practical Quantification Limit (PQL) is the name given to this MCL. Therefore, the MCL may not always be a safe level for known or suspected carcinogens, but rather the lowest amount that can be tested with reasonable accuracy. The primary reason for dispersion is thought to be the difference that subsurface heterogeneity causes between the average linear velocity and the contaminant velocities. The contaminants can travel through different flow paths to reach the same location. Groundwater moves faster in large pores than in small pores. Contaminants will therefore become less concentrated as a result of mixing with uncontaminated water. Irreversible first-order reactions and equilibrium-controlled sorption reactions are often the primary reactions taken into account when analyzing the fate and transport of pollutants in groundwater. Contaminants in a solution adhere to the solid matrix of the porous medium during sorption reactions. Adsorption and absorption are both included in sorption. Chemical species adhere to the grain surfaces during adsorption, whereas they permeate the solid grains during absorption. Contaminants will be divided between the solution and the solid phase as a result of sorption. Sorption retards the pollutants, making them travel more slowly than groundwater flow. An equilibrium reaction can be attributed to the inter-phase mass transfer process if the sorption process is rapid in comparison to the flow velocity. Specific types of biodegradation and radioactive decay make up the majority of the irreversible first-order reactions. The irreversible first-order processes, in contrast to the sorption reactions, presumably have no effect on the contaminant velocity distribution but do affect the concentration of the pollutants.

An isotherm is the relationship that exists between the concentration of the solute and the amount sorbed at a constant temperature. Because of their hydrophobic properties, many organic molecules dissolved in groundwater sorb to the solid phase. Depending on how much these compounds are drawn to the polar water molecule, these same effects cause them to become less soluble. An organic compound's tendency to partition into the solid phase increases with its hydrophobicity.

When a nonpolar molecule dissolves, the water structure changes, which triggers the sorption process of non-polar organic molecules. The linear isotherm is typically used to model this sorption process. The biodegradation of organic pollutants in groundwater is crucial because hydrocarbons and organic compounds are common contaminants. Organic contaminants undergo a biochemical process called biodegradation, which is mediated by subsurface microorganisms in the presence of an appropriate electron acceptor like oxygen. The process breaks down organic contaminants into harmless products like CO_2 and water. Anaerobic and aerobic biodegradation are the two forms of the process.

Whereas alternative electron acceptors like nitrate, Mn (IV), Fe (III), sulfate, and CO_2 are used in anaerobic biodegradation, oxygen is the electron acceptor in aerobic biodegradation. Biodegradation requires the microorganisms to break contaminants. An energy source must be available for cell growth and maintenance. A carbon source must be available in order for energy to be used to create new cells. An electron acceptor should be present along with the nutrients. A suitable environment, including salinity, pH, and temperature is required.

2.9. ANN

Developing distributed models of the fate and transport of nitrate for both soil and groundwater is difficult. In order to develop these models, the study area must be thoroughly characterized, including the physical, chemical, and biological processes. Several factors contribute to the cost and difficulty of developing these models. Many researchers have effectively employed artificial neural networks (ANNs) to overcome these challenges. An artificial neural network or ANN is made up of connections between basic processing units called neurons that can determine the relationship between input and output effects based on predefined patterns (Yoon et al., 2011). ANN solves the challenges of Physical methodologies employed in the simulation of complicated aspects of various relationships in pattern recognition. Therefore, ANN is an effective tool for demonstrating the underlying relationships controlling the input-output response patterns for various physical problems. ANN has been applied in many different hydrologic applications because of its capacity to handle complex physical problems. ANN was employed to evaluate the nitrate pollution of wells in rural areas.

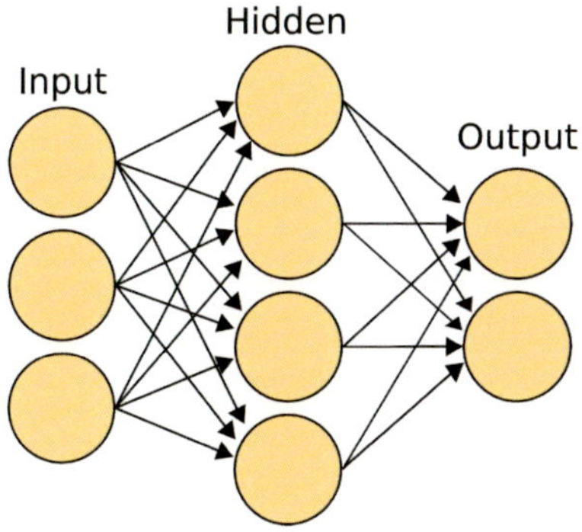

Figure 2.5. *Artificial neural networks.*

Source: https://en.m.wikipedia.org/wiki/File:Artificial_neural_network.svg, Creative Commons License.

The aquifer thickness, sampling depth, proximity to crops, streams, disposal sites, and septic systems, the topography surrounding wells, the sample collection season, and the interval between the likely dates of nitrate application and the first storms/floods were considered input characteristics. The time of year, cropping methods, water table depth, nitrogen fertilizer application rate, precipitation index, initial nitrate content, and flow were among the inputs used by the artificial neural network (ANN). They simulated how various management choices would affect nitrate leaching using artificial neural networks. Some researchers used a unique type of artificial neural networks (ANNs) (Figure

2.5) called modular neural networks (MNNs) to simulate the long-term nitrate distribution in aquifers that are dominated by agriculture.

Advanced computational units are used by MNN to do complex tasks. These tasks were created by splitting the input-output response patterns into regions of similar features, so breaking down the physical task into more manageable sub-tasks. As a result, MNN is superior to a standard multilayer ANN in its comprehension of the intricate physical situation. The distribution and methods of land use, the nitrogen sources, and the physical and chemical processes in soil and groundwater do not show high correlations with nitrate occurrences in groundwater. MNN may therefore be a useful analytical technique for simulating the long-term nitrate concentration spatial distribution in aquifers dominated by agriculture in the absence of correlations. To forecast the movement of contaminants in groundwater, analytical models are helpful tools. They can also be applied to numerical model testing.

In addition to the physical properties controlling flow in the system, a thorough grasp of all hydrologic inputs to the system is necessary to characterize an aquifer for water distribution. Placing several pumping wells in a field necessitates a thorough understanding of how long-term withdrawals will impact across the entire region. Hydraulic and transport structures must be thoroughly described for the purpose of mitigation and aquifer characterization. The goal of aquifer testing could be to better characterize organic carbon in the geologic matrix in order to comprehend potential pollutant partitioning. Geophysical investigations are employed to build a more comprehensive conceptual model. Surface geophysical methods can be used to study stratigraphy at a specific location. Standard surface geophysical techniques rely on gravity responses, electrical resistivity, or seismic reflection. These techniques can be applied to ascertain the location and existence of groundwater as well as the dip and depth of the bedrock. The observational responses obtained from these methods are usually calibrated to documented sites where geologic subsurface samples are accessible for comparative analysis. These methods can offer more spatial subsurface data, which could be helpful for hydrologic research. The signals can, however, be difficult to receive and necessitate some subjective interpretation. Two widely used surface geophysical techniques for characterization of an aquifer are electrical resistivity and seismic reflection.

2.10. AQUIFER

Aquifer profiles have been characterized by the application of seismic reflection methods (Bertoni et al., 2020). Secondary waves (S-wave) are mapped along cross-sections for characterization. Primary waves, also known as compressional waves (P-waves), move in the direction of the wave but are substantially faster. The capacity to map the depth profile between clay, bedrock, and/or unconsolidated media is made possible by the reflection of these waves from interfaces and the corresponding travel time through the geologic media. Successful high-resolution P-wave (compressional wave) seismic reflection studies have been conducted in regions with geological conditions that facilitate the acquisition of high-quality data, such as fine-grained material, shallow water table, and large seismic velocity differences. S-waves work best in unconsolidated, water-saturated sediment sequences. Because S-waves follow the sediment framework velocity, they are mostly unaffected by saturation levels and frequently give a more accurate picture of the underlying geology. Of all the geophysical techniques, downhole techniques are perhaps the most frequently employed. Since clay layers have more naturally occurring gamma radiation than other geologic materials, they can be distinguished using the natural-gamma approach. On the other hand, more permeable mediums have higher electrical conductivity, which is capitalized upon by the resistivity. There are many methods available for bore holes to characterize aquifers. The engineer must exercise due diligence by looking into and following the relevant federal, state, or municipal regulations. Monitoring well drilling is currently one of the more costly aspects of aquifer characterization, but if replacement wells need to be drilled or recently installed wells need to be abandoned, poor well design can drive up costs dramatically. To guarantee that the information gathered from an installed well is sufficient for characterization, the well must be developed properly. Aquifer characterization frequently entails gathering data on hydraulic response from monitoring wells or boreholes. Large-scale, reliable aquifer characterization can be achieved using hydraulic testing in conjunction with geophysical and lithologic records. The basic knowledge of well hydraulics and flow in porous media serves as the foundation for the hydraulic interpretations.

2.11. POLLUTANTS

Groundwater was found to contain synthetic organic pollutants (SOCs) as early as the 1970s (Postigo & Barceló, 2015). But it wasn't until the early 1980s, when regulations requiring groundwater monitoring were extended. The Safe Drinking Water Act or SDWA was amended by the US Congress in 1986 in response to the realization of the extent of groundwater contamination in the country. This amendment required the USEPA to create national primary drinking water regulations (NPDWR), which set a limit on the amount of contaminants that could be found in drinking water obtained from public water systems. The SWDA now regulates more than 50 volatile organic compounds (VOCs). Because of their extensive usage and inappropriate disposal, either through unintentional or purposeful leakage into the subsurface, these chemicals are common pollutants of groundwater. According to a US EPA survey (1990),

28% of large-scale and 17% of small-scale public drinking water treatment facilities that used groundwater as their water source had VOCs. Because of their comparatively low aqueous solubilities, the volatile organic compounds (VOCs) present in groundwater frequently exist in the subsurface in a distinct organic phase known as non-aqueous phase liquid (NAPL). Groundwater pollution can come from the long-term dissolution of NAPLs or their constituents into the aqueous phase, which can last for years or even decades. Groundwater pollution concentrations are frequently below the aqueous solubility threshold due to mass transfer limits and dilution effects. Nevertheless, the concentration of these substances usually surpasses the maximum concentration level (MCL) set forth by the USEPA. Many in situ remediation strategies have been developed over the past few decades by environmental scientists and engineers with the goal of treating soil and groundwater contamination. Federal and state laws set risk-based thresholds on contaminant concentrations within and/or emanating from waste sites have played a major role in the development of these technologies.

Petroleum-based hydrocarbons, which include heterocyclic, aromatic, alkane, and alkene components, are prevalent in groundwater. These substances are commonly discharged from refineries, underground storage tanks (USTs), and oil production facilities into the subsurface as effluents.

Figure 2.6. *Gasoline.*

Source: https://en.wikipedia.org/wiki/Gasoline#/media/File:Gasoline_in_mason_jar.jpg, Creative Commons License.

Gasoline, (Figure 2.6) a complex mixture of monoaromatic hydrocarbon chemicals such as benzene, toluene, ethylbenzene, and xylenes (BTEX), is one of the most often leaked LNAPLs underground. BTEX chemicals generally have a limited propensity to interact with the solid phase through sorption processes and are only weakly soluble in water. Because of its low organic carbon partition coefficient and relatively high aqueous solubility, benzene is comparatively mobile in groundwater systems. Even though most of the components of BTEX are prone to biotic degradation, benzene is the most resistant. Furthermore, because gas stations are so widely dispersed, they are frequently situated near residential areas and/or drinking water supplies, which make them even more important places for remedial action. Excluding petroleum contamination from underground storage tanks, there are still an estimated 30,000 to 50,000 locations in the United States with groundwater contamination in 2003 (Awal, 2009). The prevalent contaminants among the DNAPLs are chlorinated organic solvents, like trichloroethene (TCE) and tetrachloroethene (PCE), which are widely used in industrial settings. Source zones can be treated with a more resilient multi-phase extraction (MPE) method when pollutants are present in the gas, NAPL, and aqueous phases. The original purpose of MPE systems was to remediate volatile organic compounds (VOCs) in low-to-moderate permeability aquifer formations. The traditional soil vapor extraction (SVE) method, which is used in the unsaturated zone above the water table, is modified to form the basis of the MPE process. SVE systems typically entail the extraction of volatile organic compounds (VOCs) from the subsurface. MPE entails modifying conventional SVE systems to handle several phases, such as the simultaneous extraction of soil vapors and groundwater.

2.12. SURFACTANTS

Cationic surfactants are not good choices for the source zone cleanup applications due to their significant sorption tendency, high cost, and toxicity. The hydrophilic group of non-ionic surfactants is made up of polar ethylene oxide (EO) groups as opposed to an ionic species like those of anionic and cationic surfactants. The comparatively low cost, low critical micelle concentration, large solubility capacity for the majority of hydrophobic organic compounds, insensitivity to background electrolytes, and potential for subsurface biodegradation are some of the interesting characteristics of non-ionic surfactants. Most non-ionic surfactants are considered non-toxic and are used extensively in food-grade applications, pharmaceuticals, and household items: however others, like alkylphenols, exhibit toxicity to aquatic species. The possible application of non-ionic surfactants for the remediation of NAPL-contaminated aquifer formations has thus been the subject of extensive research. Surfactants tend to accumulate near interfaces because of their hydrophobic and hydrophilic structures, and they can significantly change the characteristics of liquid-liquid, air-liquid, and solid-liquid interfaces in subsurface systems. The capacity of surfactants to self-agglomerate to form micelles above a particular concentration is known as the critical micelle concentration (CMC), and it is a second significant property of these molecules. Non-ionic surfactants have a CMC of about 10–100 mg/L, but anionic surfactants typically have a CMC of more than 1000–2000 mg/L and can differ contingent on the characteristics of the system (Casero et al., 1997). Surfactants exist as single molecules or monomers at concentrations below the critical micelle concentration (CMC). The monomers start to group together to form micelles when the concentration rises above the CMC. These surfactant micelles in aqueous solutions are usually composed of a hydrophobic core that nonpolar chemical molecules readily partition into and a hydrophilic shell surrounds it. As the surfactant concentration rises above the CMC, the quantity of micelles increases while the quantity of monomers stays unchanged.

CHAPTER 3

GEOCHEMISTRY OF GROUNDWATER

Contents

3.1. INTRODUCTION

One of the most captivating substances is water. Because of its many states, mobility, and solute-transporting capacity, it plays a vital role in a wide range of geological processes. Deep crust waters carry gold and other metals to form ore deposits. Freshwater is the basis of life on land, and the water of the primordial oceans served as the incubator for the emergence of early life. Water, in its vapor form, warms the earth's climate by about thirty degrees and makes it habitable. However, when condensed into clouds, it controls our climate by reflecting solar energy. Its unusual open-tetrahedral framework, when crystallized as ice, makes it less dense than liquid water, which makes its solid form buoyant in contrast to most other compounds. Without surface ice, aquatic habitats would freeze to the bottom and seasonal albedo would not be regulated. Even the investigation of possible life on Mars is based on water evidence. The distinguished property of H_2O in geochemistry is its ability to dissolve materials and transport solutes (Glynn & Plummer, 2005). This feature is mostly caused by the asymmetrical charge distribution of the water molecule. A dipolar arrangement results from the four tetrahedrally orientated electron orbits that the two hydrogen atoms and one oxygen atom share. Despite the overall electrical neutrality, the surface charge is unevenly distributed, resulting in a net positive charge on one side and a net negative charge on the other side.

3.2. HYDROGEN BONDING

Neighboring positive and negative poles attract to create a weak hydrogen bond that can be readily broken and reformed. The end product is loose molecules that constantly disintegrate and reassemble. Approximately 80% of water molecules have this configuration at 25°C. This is somewhat lower near the boiling point, and water takes on its complete structure as it freezes. Hydrogen bonds are weak, thus when water molecules turn into water vapor, they break readily. Hydrogen bonds reform and water vapor condenses when humidity rises over the dew point, two essential processes in the hydrological cycle. Water has certain peculiar properties since it is dipolar. At Earth surface temperatures, it first permits H_2O molecules to attract one another and condense into a loosely structured/bonded liquid. Furthermore, these dipolar molecules have the ability to align themselves with ions, isolating and retaining them in solution. Water is therefore a useful solvent, a strong chemical weathering agent, and a crucial solute transport medium. The fact that the hydrogen bond between the heavier isotopes of water is stronger is another significant characteristic of water. This alters reaction rates, causing these isotopes to cleave during the hydrological cycle and providing a means of tracking the source of water. Gases are not ionized, in contrast to most solutes in water, which are present as ions. Certain gases, like carbon dioxide, combine with water to produce more unionized and ionized species that might improve solubility. Gases that are neutral in nature tend to be less soluble in water because they interact with it less. Gases from a neighboring gas phase, such as the atmosphere covering surface waters and soil water in the unsaturated zone, can dissolve in water. In these situations, the partial pressure of the gas in the nearby air controls the concentration of the gas dissolved in the water. Another way that gases in groundwater are produced is by geochemical processes in the saturated zone.

Figure 3.1. *Clay minerals.*

Source: https://en.wikipedia.org/wiki/Clay_mineral#/media/File:Kaolinite_-_USGS_bws00008.jpg, Creative Commons License.

3.3. MINERALS

Clay minerals (Figure 3.1) belong to the phyllosilicates class of minerals, which is distinguished by its structure (Bergaya & Lagaly, 2013). These are the byproduct of fundamental silicate minerals during weathering. Even though the majority of clay minerals have clay-sized particles, many of these particles could also be quartz or other elements. The structure of clay minerals, with its negatively charged borders that allow cations to be fixed and exchanged, is a common characteristic. With their propensity for cation exchange, which can impact groundwater geochemistry, they are highly significant. Aluminosilicate minerals undergo water-based modification, which results in the hydroxyl groups (OH^-) present in clay minerals. Commonly, the hydrothermal variation of mafic minerals (olivine and pyroxene) results in the production of serpentine (Figure 3.2) and chlorite (Figure 3.3).

Figure 3.2. *Serpentine.*

Source: https://commons.wikimedia.org/wiki/File:Serpentinite.JPG, Creative Commons License.

Meteoric weathering of feldspar-rich rocks typically forms kaolinite and illite, whereas meteoric modification of mafic minerals yields smectite clay. Unlike kaolinite and illite, smectites are hydroscopic and can draw water into their inter-layer/sheet locations.

Figure 3.3. *Chlorite.*

Source: https://en.wikipedia.org/wiki/Chlorite_group#/media/File:ChloriteUSGOV.jpg, Creative Commons License.

Al often replaces Si in the tetrahedral sites, while Mg, Fe, and Al substitute in the divalent sites. However, Al creates a charge imbalance that needs to be compensated. The largest carbon reservoir on Earth is found in carbonate rocks, which contain more than 100,000 times the carbon equivalent of living biomass (Einsele et al., 2001). The presence of substantial aquifers in marine limestone and dolomites can affect the geochemistry of runoff from these areas. Moreover, carbonate minerals are frequently used in sedimentary or crystalline rocks to fill up fractures/voids. The geochemistry of natural waters depends on the carbonate minerals since they can form at low temperatures. There are two main minerals, dolomite and calcite, with less common carbonates and polymorphs at higher temperatures. Because carbonates are more soluble than silicate minerals, especially in acidic environments, they will dissolve in conditions close to the surface, enhancing groundwater circulation and permeability. The highly soluble salts known as evaporite minerals are typically created when groundwater or seawater evaporates in confined basins. Even though evaporites are continuously producing, they are frequently found in specific sedimentary rocks that were bundled with shale, limestone, or dolomite and developed in Palaeozoic regions.

Figure 3.4. *Apatite.*

Source: https://commons.wikimedia.org/wiki/File:Apatite_Canada.jpg, Creative Commons License.

In aqueous systems, phosphate (PO_4) is a minor species, but it frequently serves as the nutrient that limits photosynthetic activity. The phosphate in apatite, (Figure 3.4) a mineral found in many igneous rocks, is the major repository of phosphorus. There are four types, based on the interchangeable anion. Complex calcium phosphates precipitate as marine deposits in the form of phosphorite formations in sedimentary rocks. Iron in lakes and rivers can also complicate phosphate to form low-solubility ferrous and ferric phosphates. Sulfide (S^{2-}) is the reduced form of sulfur found in the environment. Most transition metals can easily form sulfide (S^{2-}), which is present as hydrogen sulfide gas (H_2S) or ionized as HS^-. Pyrite, or iron sulfide, is the most prevalent sulfide mineral. Mostly found in waste rock and mine tailings, it is an unprofitable mineral in many mining operations. In organic-rich shales and certain limestones, pyrite is present. There is frequently a small amount of pyrite visible in glacial till due to bedrock erosion. In low-temperature aqueous environments, the more amorphous iron sulfide mineral mackinawite can develop. Iron sulfides have the ability to oxidize, which makes them key regulators of sulfur, and iron geochemistry in natural waters. When pyrite and mafic minerals weathers, iron and manganese hydroxides and oxides are formed. They have the ability to crystallize into a specific mineral or produce hydrated amorphous deposits on organic or mineral surfaces in soils. They are present as minerals that create rocks, especially metamorphic rocks. They play an important role in soils as negatively charged substrates for the adsorption of metals and pollutants, as well as redox buffers. The various sizes and compositions of aquifers and aquitards have an impact on the geochemistry of groundwaters. Aquifers are technically described as water-saturated geological mediums that may store and transfer groundwater (Narasimhan, 1998). They are often made of fine-to coarse-grained porous media or karst, or fractured bedrock. Aquitards are hydrogeological units that do not easily transfer water despite having a high-water content. Aquitards are often low-permeability materials like shales and sediments that are high in clay. The thinnest hydrostratigraphic units are unconfined or phreatic aquifers, which are made up of an unsaturated zone that is above the water table and a saturated zone that extends upward to the water table. Unconfined aquifers are typically replenished by infiltration from nearby surface water bodies or by meteoric waters passing through the surrounding soil and unsaturated zone. Through the unsaturated zone, soil CO_2 and atmospheric gases exchange with groundwater. Lower-permeability aquitards lie on top of confined or artesian aquifers. Recharge occurs through either groundwater leakage from the nearby confined aquitard or flow from an area where the aquifer experiences unconfined conditions.

3.4. AQUIFERS

In addition to sandstone and other lithified strata having intergranular porosity, porous aquifers are generally unconsolidated sands and gravels that were deposited by glaciers or rivers. Porosity, or the volume of pore space, can range from 20% to 30%. The most stable of the major clastic minerals, quartz, usually predominates in sand and sandstone aquifers, with some feldspar. The weathering of parent materials and the movement of clastic sediments have dissolved important minerals into clays. There may also include heavy minerals including garnet, zircon, and magnetite. Although they are frequently found in secondary mineralization, carbonate minerals are less common. Ferrihydrite, which is produced when ferrous iron-bearing minerals oxidize, is a common example of other secondary mineralization. Unconsolidated aquifers may contain buried sedimentary organic matter due to their sedimentary origin, which has significant implications for groundwater redox reactions. This is especially true for clay-rich aquitards, which are usually formed in marine or freshwater environments with a large organic sedimentary material. The majority of fractured rock aquifers are found in crystalline silicate rocks, which include volcanic, metamorphic, and plutonic rocks. Aquifers with this type of porosity are created when sedimentary rocks like sandstones and shales crack. Sandstones frequently exhibit secondary intergranular porosity. Microfractures and intergranular discontinuities, which are secondary porosities in unbroken granites and metamorphic rocks, are frequently used as sources of solutes rather than as substantial water storage. Water and other solutes move through this porosity only by diffusion; however groundwater passing through the open cracks may get salinized from these sources. Because silicate minerals are less soluble, tectonic activity causes fracture development. Fractured aquifers always have low bulk porosity, despite large variations in fracture density and pore. Feldspar is the predominant mineral in fractured silicate aquifers. Alkali feldspars, or potassium and sodium feldspars, are the building blocks of granites and are the sources of significant quantities of Na and K in groundwater due to weathering.

Mg-Fe minerals and calcium feldspars predominate in rocks with mafic composition (Wu et al., 2017). Because of their intense fracturing from joints and along flow beds, basalts are a great source of water. They are mostly composed of plagioclase feldspar (Ca & Na), with extra pyroxene and amphibole. One of the freshwater resources that are heavily utilized worldwide is karst aquifers. The main mineral, calcite, dissolves in acidic environments. As a result, fissure dissolution causes porosity. Despite being less soluble, dolomite nevertheless contributes to favorable aquifer conditions. The higher limestone and dolostone surface have fissures, as groundwater replenishing through soils gains carbonic acid due to CO_2 dissolving in the soil. As a result, dolostone and limestone aquifers frequently exhibit phreatic behavior. The level of detail in water-quality assessments varies, ranging from a basic examination of the main species to a more extensive examination that takes into account isotopes, minor and trace metals, dissolved gases, organic carbon, bacteria, pollutants, and radioactivity. Isotopes and a simple geochemical analysis may be all that is needed for studies on the dynamics of watersheds and the management of water resources in order to identify the source of salinity and groundwater recharge. An extensive list of analytes would be needed for drinking-water quality monitoring as well as in-depth research projects like geochemical exploration and the characterization of contaminated areas. To ascertain the speciation or form of the dissolved component, more intricate investigations incorporating selective preservation techniques and filtering at various pore sizes might be conducted. Radiological parameters are frequently measured' these are expressed as the activities of specific radionuclides or as alpha and beta activity. The alkali metals Na^+ and K^+ as well as the alkaline-earth Ca^{2+} and Mg^{2+} are the main cations found in most meteoric fluids. These elements are the main cations of the minerals that create rocks, such as the carbonates (calcite and dolomite) and silicates (feldspar, micas, amphiboles, and pyroxenes). The primary anions are the carbonate species, which are produced by soil CO_2 weathering and include HCO^{3-} (bicarbonate) and CO_3^{2-} (dissolved carbonate). Both sulfate (SO_4^{2-}) and chloride (Cl^-), which are highly soluble ions found in crustal fluids, salt deposits, and seawater, are byproducts of weathering of sulfide minerals and evaporite beds, or from marine sources.

3.5. GEOCHEMISTRY

The primary geochemical species make up more than 95% of the total dissolved solids or TDS in natural rivers/water bodies (Huang et al.,2009). Anthropogenic sources can also be found in agricultural, urban, and industrial activities. The water isotopes, D and $_{18}O$ offer opposing views on the water formation. These isotopes are now frequently included in geochemical assessments. The isotopic composition of dissolved organic carbon or DOC and dissolved inorganic carbon limits the source of DIC and reveals information about the biogeochemical history of water. Due to their low crustal abundance and low solubility under oxidizing conditions, the heavy and transition metals are usually found in waters at trace (ppb) amounts. Redox can improve transport of U in oxidizing conditions or Fe in reducing conditions. Redox is a significant solubility regulator. Depending on the goals of the investigation, the trace metal analysis might range from a partial scan of all metals to a thorough analysis of a subset of analytes. Dissolved gases, such as O_2, CH_4, H_2S, and CO_2, represented as ppm, are a common component of geochemical investigations and are crucial for tracing contamination and redox processes. Methane is a sign of extremely anaerobic circumstances, whereas dissolved oxygen indicates aerobic conditions. H_2S is extremely toxic and can be fatal, but CH_4 can build up in homes and pose a fire hazard. In order to reconstruct groundwater recharge conditions, very detailed investigations may evaluate the concentration of the noble gases or of the primary atmospheric gases O_2, N_2, and Ar. An independent gas phase for the investigation of the principal gases as molar or percent fractions can be obtained from groundwaters. It is also possible to monitor the concentrations of dissolved gases, which enables thermodynamic estimates of subsurface reactions.

Isotopes can be used as a helpful tool to determine the source of solutes and water. The distribution of isotopes in some systems is governed by the interaction of two processes: isotope fractionation during physical or chemical reactions, and isotope distillation from reactant reservoirs as the reaction moves forward. These two procedures divide isotopes into various types. When conditions deviate from equilibrium, as when a reactant is added or the pH is changed, physicochemical reactions are accelerated.

Figure 3.5. *Calcite.*

Source: https://commons.wikimedia.org/wiki/File:Calcite-67881.jpg, Creative Commons License.

When calcite (Figure 3.5) is precipitated from water through a sudden increase in ionic strength, the resulting calcite precipitate has the same δ 13C $CaCO_3$ as the original δ 13 CDIC (Samanta et al., 2015). Similar to this, redox reactions rarely or never have a reversible reaction. But the majority of redox processes occurring in close proximity to the surface are mediated by bacteria and archaea, which exhibit remarkable selectivity in the isotope bonds they choose to split. Using the energy involved with the exchange of electrons between redox-sensitive elements, such as carbon, sulfur, nitrogen, hydrogen, and iron, microbes colonize a wide range of geochemical conditions, from acidic mine tailings to hot springs and hypersaline conditions. In order to increase the energy, they obtain from these reactions, bacteria and archaea can choose light isotopes that have weak bonds, such as $_{32}$S–O during sulfate reduction rather than $_{34}$S–O. The sulfide produced by sulfate-reducing bacteria is decreased in $_{34}$S compared to SO_4^{2-}.

The corresponding enrichment factors in microbially mediated processes are not always adequately limited to temperature–fractionation proportionality. In addition to temperature, additional factors that might impact fractionation include the concentrations of reactants, the availability of related nutrients, the concentrations of other redox species that can influence the process, like O_2, and the strains of bacteria that are participating in the reactions. Another process that can fractionate isotopes is diffusion, which is theoretically connected to the relative molecular mobility of a heavy isotope species in comparison to a light isotope species. It is a crucial mechanism in aquitards, where molecular diffusion in pores dominates transport and groundwater advection is negligible. Many geochemical processes occur in non-equilibrium (kinetic) environments. Isotope exchange will occur at the mineral surface if mineral-solution reactions occur under equilibrium, that is, with equal forward and reverse reaction rates, or during growth. However, the reaction is quantitative during net dissolution because it is unable to select isotopes below the surface. In these kinds of quantitative reactions, the reactant isotopic composition is maintained in the reaction products. The hydrological cycle provides more than 1000 km^3 of freshwater per day. More than 90% of water returns to the troposphere. Approximately 75 km^3 of freshwater is released into the ocean every day as a result of the remaining water runoff over the surface or penetrating into soils and aquifers. The monitoring of surface water outflow is crucial for managing water resources, producing hydroelectric power, and preventing flooding. In order to track water levels, recharge rates, and water quality, test wells are drilled into aquifers to monitor groundwater supplies. The interpretation of physical factors, such as variations in groundwater level, in conjunction with data on temperature and rainfall, is essential to the assessment of water resources. An additional tool for differentiating between various water sources and recharge zones is provided by isotopes. The natural isotopes $_{18}O$ and D are excellent tracers because they are an integral part of the water molecule. Every stage of the hydrological cycle, from initial evaporation over the oceans to condensation, precipitation, groundwater recharge, and runoff back to the seas, is where they are selectively divided.

3.6. ISOTOPES

The largest source of water vapor is the evaporation of ocean water, with other quantities coming from surface water evaporation and transpiration from vegetation. The cooling of the vapor cloud produces precipitation. For condensation to occur, the vapor temperature needs to fall below the dew point, which is the point at which humidity equals 100%. Warm air rises in convective precipitation systems, orographic rainfall over terrain, or displacement over cold fronts, causing adiabatic expansion and cooling. This is known as a rainout, and it occurs when the temperature drops. Rain and snow contain $_{18}O$ and D, which are primarily partitioned into the vapor phase during rainout due to fractionation between vapor and water or vapor and ice within the cloud (Ingraham, 1998).

The isotope evolution will proceed according to a Rayleigh-type distillation if there are no additions to the vapor mass. Rainout progresses toward colder, isotopically depleted precipitation as a result of the remaining vapor in $_{18}O$ and D being gradually depleted in following rains. The primary mechanism for dividing $_{18}O$ and D is Rayleigh distillation, although recycling of precipitation through transpiration and evaporation also has a substantial impact on the isotopic composition. As vapor masses migrate inland, their closeness to marine waters influences the isotope composition of precipitation. Conrad's index, which increases with latitude and distance from the coast, represents continentality and is dependent on the amplitude of the seasonal temperature variation. When it comes to mean annual precipitation, stations with higher continentality have lower $\delta\ _{18}O$ and δ D values. In the high troposphere, continentality is linked to increase mixing with isotopically more depleted vapor masses. Rivers that flow from continental scale basins frequently exhibit it in the form of a depleted isotopic composition in relation to local precipitation in the downstream regions.

The tropospheric lapse rate, which measures the temperature drop over 100 meters, is approximately 0.6°C. Precipitation should then exhibit a 0.4% decrease in $_{18}O$ for every 100 meters of elevation gain (Dutton et al., 2005). Actually, for every 100 meters of altitude gain, the $\delta\ _{18}O$ of precipitation drops by 0.15 to 0.5, while δ D drops by 1 to 4 meters. Because temperature gradients are frequently less in warmer places, the altitude effect for $\delta\ _{18}O$ per 100 m elevation is typically less than –0.2%. Based on long-term weighted average of precipitation, lower elevation catchment with relief contrasts as low as a few hundred meters likewise show an altitude effect for $_{18}O$ and D. Since that groundwater recharge is frequently a long-term weighted average of the precipitation in the recharge area, it is better to monitor this in a research location. In low-relief catchments, recharge elevations can thus be identified. Relatively stable isotope concentrations in precipitation can be used to track climate change.

Glacier ice cores include longer records of precipitation. These cores offer a remarkable temperature change record that goes back over 650,000 years based on the differences in $\delta\ _{18}O$ and δ D. Some sources of these climatic records can be discovered in the gradual build-up of speleothem from groundwater, where the $\delta\ _{18}O$ of the groundwater that generated the deposit can be reconstructed using the $_{18}O$ in calcite ($CaCO_3$). The record from Devils Hole, Nevada, where local groundwaters flowing through an exposed fault have coated the wall with up to a meter of carbonate, is one example. A 500,000-year climate record that documents the start and length of glacial episodes was obtained from profiles of $\delta\ _{18}O$ along cores from this speleothem (McDermott, 2004).

An isotope record of the event indicated by depletion in $_{18}$O and D is found in groundwaters replenished during these glacial events or by meltwater at the conclusion of continental glaciation. Approximately 12,000 years ago, during the Pleistocene, the Laurentide Ice Sheet melted, causing a substantial hydrological and morphological influence on the majority of the North American region. The deep circulation of glacial meltwater into bedrock aquifers generated by the severe hydraulic gradients near the edge of the ablating ice sheet is where remnants of the meltwater event may still be found today. Glacial meltwater remained in place with minimal flooding because of the degraded slopes in these aquifers when the ice receded. Since the Pleistocene, high latitude regions have seen significant temperature fluctuations.

These include the cold conditions of the last glacial maximum or LGM, which occurred 20,000 years ago, the early Holocene warming that resulted in substantial permafrost degradation, the late Holocene cooling that culminated in the Little Ice Age, which lasted from 1400 to 1800, and the warming that followed in the twentieth century (Kjaer et al., 2022). Thaw slumps occurred when permafrost became exposed at the surface. One method for reconstructing the Quaternary past at high latitudes and understanding the genesis of ice in permafrost is the use of stable isotopes.

Although carbon is the fourth abundant element in the universe, its quantity in the crust is only 200 parts per million. Biological cycling of carbon nevertheless leaves its geochemical trace on the hydrosphere through cell and energy conversions. Its ability to form organic molecules and store energy is remarkable due to its diversity of redox states. From a geochemical standpoint, carbon, both organic and inorganic, is significant. Carbon that is inorganic or mineralized comprises CO_2 and its hydrated forms, such as H_2CO_3, CO^{3-}, and CO_3^{2-}. Major types of organic carbon include carbohydrate (CH_2O), which is the major component of all biomass on land and in the ocean, and more advanced compounds like hydrocarbons and humic substances. Both CO_2 as dissolved inorganic carbon (DIC) and dissolved organic carbon (DOC) in groundwaters are essential to weathering processes and the groundwater geochemistry development.

3.7. PRIMARY PRODUCTION

Through gross primary production (GPP), terrestrial forests and grasslands sequester 122 GtC, or 17% of atmospheric CO_2, each year. Over 650 GtC of biomass is found on the Earth today. A further 1500 GtC is present in soils as soil organic carbon (SOC), which can be used by bacteria for respiration, returning the majority of the CO_2 to the environment. In response to increased CO_2 fertilization, GPP and terrestrial biomass are both rising, especially in agricultural areas and boreal forests. Thus, a major factor regulating atmospheric CO_2 concentrations is the net flow between CO_2 uptake by GPP and return to the atmosphere through respiration. A large portion of the annual variability in atmospheric CO_2 can be attributed to variations in soil temperature and humidity.

There is a gap in the correlation between this net flux and temperature, indicating that temperature may contribute to some of the atmospheric CO_2 increase that has occurred over the past century. This is true on a seasonal basis, since seasonal differences in respiration and photosynthesis influence monthly atmospheric CO_2 concentrations, giving rise to the concept of Gaia. Marine habitats serve as a substantial carbon sink. Approximately 28 ppm of DIC, primarily in the form of HCO^{3-}, is found in sea surface waters (Dhanu Radha et al., 2022). Marine waters and atmospheric PCO_2 are almost in balance (Henry constant). The 900–1000 GtC found in the shallow waters are primarily found as DIC, with another 3 GtC found in algae and DOC.

Approximately 90% of the shallow ocean reservoir, or 90 GtC, is exchanged with the atmosphere each year. The deep ocean has around 38,000 Gt of carbon. One of the geological restraints on the carbon cycle is the annual release of roughly 0.2 GtC of mantle CO_2 into the atmosphere by terrestrial volcanoes. Similar amounts of DIC are eliminated by continental weathering and flow into the ocean through aquifers. From the standpoint of groundwater geochemistry, this flux is significant even though it is small in relation to the biological cycle of carbon.

When taking into account the CO_2 fluxes from both land and sea, more than 25% of the atmospheric CO_2 (210 GtC) is exchanged each year, meaning that CO_2 has a very short atmospheric residence period (Starr, 1993). The annual addition of approximately 6 GtC, or 3%, to this total flux comes from the combustion of fossil fuels and cement manufacturing. Because atmospheric CO_2 has a short residence time, the anthropogenic input is blending with the combined stores of soil, marine, and atmospheric carbon, totaling around 4100 GtC, or 0.1% yearly. Because of this, the yearly rise in atmospheric CO_2 of roughly 1.7 ppmv, or 0.4%, is smaller than the 0.8% increase estimated from merely mixing 800 Gt of atmospheric CO_2 with 6 Gt of water.

3.8. WEATHERING

Acidity substantially aids in the chemical weathering of silicate and carbonate minerals. The majority of minerals are rarely soluble in clean water. Carbonate minerals, like dolomite and calcite, have easy dissociation processes and low dissociation constants, resulting in the production of Ca^{2+} and Mg^{2+}, and CO^{3-}. But adding acid accelerates the melting of carbonate. Aluminosilicate minerals like feldspar weather due to hydrolysis processes that take up hydrogen ions and increase pH. As a result, adding acid makes these minerals more soluble, which considerably speeds up weathering. The primary source of the acidity needed for weathering in natural soils is the oxidation of organic matter, which releases carbonic acid (H_2CO_3), which is created by hydrating CO_2. While CO_2 via SOC respiration reaches 3,000–10,000 ppmv, atmospheric CO_2 concentrations are only 400 ppmv (Faimon & Lang, 2018). Before terrestrial plants and soils arrived, geochemical weathering was facilitated by CO_2 concentrations in the atmosphere that were higher than those of the present day. The formation of the terrestrial biosphere and the lowering of atmospheric CO_2 concentrations limited continental weathering to the soil horizon that covers bedrock. When carbonic acid can be replaced as it is consumed, the weathering effect of carbonic acid is significantly increased. This is true in unsaturated soils, where mineral weathering occurs in the soil water in direct contact with soil air that is rich in carbon dioxide. As weathering consumes carbonic acid in the soil water, CO_2 from the soil air replenishes it. Groundwater does not come into direct contact with the soil atmosphere in the saturated zone beneath the water table. Rapid changes in the environment lead to elevated pH levels and decreased water-rock contact.

The breakdown of organic materials releases nitrogen and sulfur, which are in reduced forms in organic compounds, in addition to CO_2. When these elements mineralize into substances like sulfide (H_2S) and ammonia (NH^{4+}), they become accessible for oxidation in aerobic reactions that are often carried out by bacteria. Reduced chemicals that undergo oxidation to release acidity can also be found in bedrock. Sulfide minerals, like pyrite, can be a part of unconsolidated tills and are frequently found in igneous rocks, limestones, and shales. These minerals are subject to bacterial oxidation when soils grow. Weathering releases ferrous iron (Fe^{2+}), which is found in pyrite and many other ferromagnesian silicates like olivine (Fe_2SiO_4) and pyroxene ($FeSi_2O_5$). Oxidation of the iron releases more acidity. This source of acidity is demonstrated by the precipitation of ferric iron hydroxide ($Fe[OH]_3$) in soils. The process of weathering carbonate rocks occurs faster than that of silicate rocks. Carbonate minerals can be found in many different forms, such as soil crusts and cements, as well as secondary phases that fill in fractures in aluminosilicate rocks and limestone and dolomite formations. Whether in crystalline rock fractures or sedimentary rocks, the breakdown of carbonate creates porosity that promotes groundwater advection. Carbonate bedrock, limestone, and dolomite are crucial for providing groundwater to wells and surface water bodies. Carbonates have a dominant effect on the geochemistry of watersheds due to their fast rate of weathering, despite their low relative abundance i.e., less than 0.1% of the lithosphere. The weathering of carbonate rocks is the cause of the high Ca^{2+} and Mg^{2+} bicarbonate content found in hard water. Weathering of the carbonate minerals lining fractures can still compensate for silicate weathering even in crystalline terrains. A large portion of carbonate weathering takes place in open systems, where organic matter and soil CO_2 reserves replace the carbonic acid used in mineral dissolution. A steady PCO_2 and consistent carbonic acid content are maintained through contact with soil CO_2. Under these conditions, carbonate weathering continues until the system approaches calcite saturation, at which point there is no more net dissolution. Under closed conditions, carbonate dissolution is prevented from widespread weathering by the restricted CO_2 acquired during infiltration via the soil. The PCO_2 will decrease with high pH in this condition, frequently without reaching calcite saturation. The partial pressure of the soil CO_2 eventually controls the amount of carbonate dissolved in an open environment (Karberg et al., 2005).

As a variety of stable aluminosilicate clay minerals are formed, these high-temperature minerals weather, releasing cations and silica in the process. All silicate reactions are acid-consuming, and they are noticeable in soils with high carbonic acid and PCO_2 when operating in an open system. Thermodynamics only favors the precipitation of feldspars or other primary silicate minerals at low temperatures when the pH and cation activity are high. The amount of available acid typically acts as the only regulator over silicate weathering in the low salinity, freshwater environments seen in most groundwater. This is not the case with carbonate weathering, where the pH and Ca^{2+} and HCO^{3-}concentrations become stable after a calcite solubility limit is reached. However, carbonate minerals such as Ca^{2+}, Mg^{2+}, and Fe^{2+} can occur in areas where silicate weathering releases divalent cations, limiting the accumulation of these cations and HCO^{3-}. Despite having a high solubility, amorphous silica readings typically fall between 10 and 30 ppm, rarely exceeding 50 ppm. This is caused by both sorption and the slow rate of weathering of silicate minerals, especially in closed systems. Ion ratios can be used to differentiate between silicate and carbonate weathering. The formation of alkali elements, especially Na^+, can be a sign of feldspar weathering. It needs to be adjusted for chloride salinity, though, by deducting the moles of Na^+ that come from marine and halite sources. Runoff from carbonate watershed weathering is enriched in alkali earths rather than alkali cations, with Ca/Na molar ratios higher than 10. Alkalinity is produced when acidity is consumed during mineral weathering. Carbonate alkalinity is the main buffering agent because carbonic acid is the main weathering agent in the natural groundwaters. Alkaline species have the ability to bind with H^+ and form an acid buffer in water, such as HCO^{3-}. For instance, this is crucial in preventing acidification in rivers that get acid rain or runoff. The primary anions responsible for this acid-buffering ability are HCO^{3-} and CO_3^{2-}. In certain groundwater, dissociated weak acids such as hydrogen sulfide (HS^-), borate (H_2BO^{3-}), silicate (H_3SiO^{4-}), and even hydroxide (OH^-)

may also have a role. The salts of some bases, such as portlandite ($Ca[OH]_2$) or the evaporite mineral nahcolite ($NaHCO_3$), can contribute significantly to alkalinity in certain rare geochemical environments.

3.9. LEACHING

The leaching of base cations and silica, combined with the low mechanical erosion enriches the regolith with iron and aluminum hydroxides. When exposed to a long dry season and a seasonal monsoon climate, these hydroxides become oxides and oxyhydroxides. Thus, iron oxides and hydroxides, such as ferrihydrite ($Fe[OH]_3$), goethite (FeOOH), hematite (Fe_2O_3) with bauxite ore of gibbsite ($Al[OH]_3$) and diaspora (AlOOH), dominate the mineralogy of these laterite formations. Because silica is more soluble than iron and aluminum, it is then preferentially removed during the extensive weathering of bedrock. After weathering over millions of years, Fe and Al oxyhydroxide minerals start to dominate laterites. The aluminum-to-iron ratio in laterites gradually rises because gibbsite is even less soluble than ferric iron hydroxide, producing bauxite ore that is appropriate for the smelting of aluminum. Many tropical laterites or ferricretes have undergone a redox step during their formation. This suggests that the iron is not solely produced by relative enrichment through silicate leaching. Rather, it is believed that iron is added locally through transportation of ferrous iron under reducing conditions. This is a crucial factor to take into account when studying the evolution of landscapes, since the accumulated ferric oxyhydroxides are a good indicator of chemical weathering. Natural groundwater geochemistry develops outside of the weathering environment via diverse mechanisms such as cation exchange, salt dissolution, and mixing with brines and other groundwaters. The soluble minerals are usually leached from soils and the shallow subsurface, leaving behind mineral products as clays and iron oxyhydroxides. Coastal areas and deeper geological situations are more likely to have higher salinity groundwater.

There are two ways to think about how surface processes affect the geochemistry of groundwater. One is the change in the primary cation composition (water) caused by cation exchange with the aquifer solid surface. Second, trace organics and metals are adsorbed from a solution onto a solid surface and desorb from a surface back into the solution. Solid surface reactivity increases sorption and exchange processes. As a result, materials with the smallest grain sizes have the major surface areas and are therefore the most reactive during the processes of sorption and exchange. The surfaces of non-crystalline substances like organic matter, colloidal particles, and amorphous ferric iron oxyhydroxides have higher electrostatic charges. The pH of the pore fluids influences a net surface charge (clay) and, consequently, its exchange capacity because hydrogen ions, or H+, can adsorb and desorb from surface charge sites. The point of zero charge (PZC) of a surface is the pH at which it's overall charge is zero (Bakatula et al., 2018). The low PZC of Si, which permits a negative surface charge for pH levels above roughly 3, is the source of the high CECs for which silica clay minerals are known. Only at neutral to high pH levels do iron oxides and hydroxides have a negative surface charge because PZC occurs in the pH range of roughly 5 to 9. At pH 9, gibbsite, $Al(OH)_3$, has an even higher PZC. On the fracture surfaces of primary silicate rocks, where weathering takes place, clay minerals are found. Sands rich in feldspar and glacial till are examples of clastic material that weathers into clays.

3.10. COLLOIDS

Colloids are non-crystalline particles (less than 10 μm) that can be either organic or mineral-based and are dispersed in water (Krause et al., 2020). Mineral colloids are usually oxides of Si, Fe, Mn, or Al found in most natural fluids. They have an extremely high, negative surface charge due to their small size, large surface area, and non-crystalline structure. Electrostatic repulsion prevents flocculation in freshwater environments, but it is inhibited in salt liquids, like seawater, causing colloids to come out (settle) of solution and flocculate. Advection is one method of moving colloids that are suspended in water. Colloids can also be created by dissolved organic molecules in natural fluids. Waters with a brown color are usually rich in humic colloids, which exhibit the property of colloids to absorb and disperse specific light wavelengths. These colloids have similar adsorption and chelation capabilities to their inorganic counterparts because of the strong negative charge that is present on the ends/terminus of many dissolved organic molecules. Heavy metals and other pollutants can adsorb onto their surfaces due to their strong negative surface charge. Colloids offer a path for the advective transfer of contaminants in surface waters and groundwaters. Zeolites are a type of low-temperature tectosilicates, similar to feldspars, that have an open crystal structure made of interconnected silica. They result from the weathering of volcanic rocks and sediments rich in feldspar in environments with high salinity and pH. They are able to incorporate significant volumes of hydration water due to their porous structure. Hydroxyl ions coupled with Si, Al, or Fe cations provide a negative electrostatic potential on the surface, which is a major source of surface reactivity. Sorption of protons can satisfy these electric potentials. But as ionic strength rises, additional cations in the solution compete with one another for these exchange sites. Protons and cations can be displaced by other cations in solution that may have a higher concentration or a greater electrostatic charge since they are only kept together by weak electrostatic bonds.

CHAPTER 4

MOVEMENT OF PESTICIDES

Contents

4.1. INTRODUCTION

Pesticides have been and will be a crucial part of protecting agricultural products. However, there are drawbacks to using pesticides for good purposes. Since around 50% of Americans rely on groundwater for drinking water, and that number may reach 90% or higher in rural areas, pesticide contamination of groundwater has raised significant concerns in recent years (Srivastav, 2020). The majority of pesticides are either accidentally spilled or handled improperly (point sources) or released into the environment as a result of agricultural application (non-point source/ point source). These substances may eventually find their way into the soil, where they may undergo microbial breakdown, volatilization, chemical transformation, and adsorption-desorption on soil elements. Pesticides that are make it through the unsaturated zone and into ground water are resistant to microbial and chemical breakdown processes. The physicochemical nature of the chemical, the flow characteristics, and the features of the unsaturated saturated zone are some of the elements that control the rate and amount of pesticide that is transmitted from the land surface to groundwater. Furthermore, pesticides have the ability to enter ground water directly through both natural and artificial voids, such as macropores and drainage wells. It is important to evaluate the possibility of agricultural applications contaminating groundwater. It is crucial to develop the best management practices that reduce potential effects. We need to reduce the contamination brought on by unintentional pesticide spills. In relation to pesticide migration, the water balance of the system should be taken into account along with the temporal and spatial variations in topography and climate, as well as the moisture conditions i.e., whether or not the soil has been saturated by previous rainfall events and the timing of pesticide application. The biological degradation would be encouraged by warm, moist conditions both during and after pesticide application. There is a chance that a heavy downpour just after application will result in pesticide loss via runoff. The likelihood of leaching is increased by relatively dry conditions, pesticide application, and less intense rainfall episodes. Volatilization rather than drainage or leaching is more likely to occur after pesticide application in hot and dry conditions (Kladivko et al., 2001).

4.2. PESTICIDES

There are several ways that pesticides can move through soil, such as by migrating through dissolved or suspended water, adsorbed soil particles, or vaporized soil air. Pesticides can travel in large quantities in suspended or dissolved forms. Water responds to a number of forces, including capillary and gravitational forces, to cause mass flow. Runoff is the word used to describe the movement of surface water causing pesticides to leave their initial location.

When water input exceeds infiltration, runoff occurs. Losses from the dissolved and sediment-adsorbed pesticide are included in pesticide runoff. Even though runoff typically causes surface water contamination directly, it can indirectly cause groundwater contamination by replenishing groundwater with surface water. Climate, site topography, soil structure and texture, pesticide characteristics, and management have an impact on runoff losses. The timing of rainfall in relation to the application of pesticides is especially significant because, in the runoff-active zone, the processes of volatilization, photolysis, leaching, and decomposition cause the pesticides to rapidly diminish over time. The main method that pesticides with solubilities less than 2 ppm are removed is through suspended soil particles. Strongly adsorbed pesticides will be held at the soil surface and particularly vulnerable to runoff. Raindrop impact fractures soil aggregates and loosens soil particles. One of the most promising best management practices for preventing soil erosion below acceptable levels may be conservative tillage. With the implementation of conservative tillage, it is generally believed that discharge of agricultural pesticides will be significantly decreased. But there is a chance that these techniques will include more herbicide application. Many types of tillage systems may be site- or chemical-specific in their ability to reduce runoff or erosion. The way pesticides partition between vapor, soil solution, and soil particles is the primary factor influencing their destiny in soil-water systems. In the unsaturated-saturated zone, their mobility is determined by partitioning and solubility. Pesticide partitioning between soil phases is contingent upon the physical-chemical characteristics of the target soil medium and organic component. The compound solubility and the adsorption-desorption processes define how the phases are divided into solution and solid. The volatilization process regulates the transition between the solution or solid and gaseous phases. The primary factor preventing pesticides from seeping into groundwater is soil adsorptivity. Unadsorbed molecules in the soil have a tendency to travel with water and can contaminate groundwater. Pesticide molecules move from mobile phases, such as liquid or vapor, to stationary phases, such as soil particles, by a process called adsorption' the opposite process is called desorption. Pesticide molecules have the ability to travel in the vapor state in the unsaturated zone. Volatilization can cause pesticide losses that can exceed 80–90% of the total amount applied (Spencer & Cliath, 1973). Although the volatilization process may not directly impact groundwater contamination, it can have a significant impact on how pesticides end up in the ecosystem.

Pesticides that find their way into the soil and water undergo both biotic and abiotic changes. As a result, mobility by itself is not a reliable measure of the possibility of groundwater pollution. During a compound retention time in the unsaturated zone above ground water, its ability to degrade into an inert/toxic form depends on its mobility, persistence, and degradation pathway. The likelihood of short-lived chemicals contaminating groundwater is significantly lower than that of persistent ones. It should be mentioned that there are other pathways for deterioration and that the quality of ground water may also be impacted by derived products. Pesticides can undergo chemical changes as adsorbed forms on solid phase surfaces or as solutes in the aqueous phase. In the field, hydrolysis, redox reactions, and other processes like photolysis are examples of naturally occurring chemical changes. The most common processes in the water and sediments are hydrolysis and redox reactions. The term hydrolysis describes the splitting of a pesticide molecular bond and the subsequent creation of a new bond with an oxygen atom.

The majority of abiotic pesticide oxidation takes place in surface waters, while abiotic reduction takes place in bottom sediments and anaerobic water. Chemicals exposed to sunlight on the land surface can undergo both direct and indirect phototransformations through the excited state reaction with other chemicals. Hydrolysis, substitution of acids, bases, elimination, and ring cleavage are the primary chemical processes that change hydrophobic and hydrophilic pesticides in an oxidizing environment. Temperature, ionic strength, and pH all have an impact on these processes. In terrestrial and aquatic habitats, the main mechanism for pesticide detoxification is degradation through microbial activity. Pesticide compounds can be completely broken down by microbiological action. Pesticides undergo three main microbial transformations: biodegradation, polymerization, and accumulation.

Pesticides can either be co-metabolized by microbes through microbial metabolic processes during biodegradation, or they can be used as substrates by the bacteria for energy and nutrients. Pesticide-soil interactions, such as adsorption and partition, have a significant impact on biological activity. The majority of clay minerals and soil organic matter are ionically adsorbed to cationic herbicides. In neutral or alkaline conditions, adsorption is higher and bioactivity is lower. Other variables that influence microbial activities, like temperature, moisture, nutrient content, and pH, all have an impact on the biodegradation of pesticides. For instance, lowering soil moisture generally tends to slow down the pace at which pesticides degrade. pH can impact not only the microbial population but also the adsorption of ionic pesticides (Liu et al., 2018).

4.3. PESTICIDE DEGRADATION

Re-application of a pesticide to the same location can occasionally result in accelerated degradation, a situation where the pesticide breaks down faster than usual. One potential solution to the current physicochemical method for treating pesticidal contamination could involve the use of microbes and enzymes. Nonpoint sources are typically the result of farming techniques that apply pesticides over large areas. Effluents in hydrogeologically vulnerable areas or close to man-made conduit systems have raised concerns. Prior to making decisions about cleaning procedures, the health and environmental hazards posed by the pesticide contamination must be assessed.

The characteristics of the chemicals and soil elements that pesticides interact with, the hydrogeology of the region and meteorological conditions all influence how pesticides behave in the environment. The chemical characteristics of pesticides have been examined in relation to traditional chemical groups, including carbamates, triazine, phenylureas, and others. Properties like the molecular formula, physical state, decomposition temperature, solubility in different organic solvents, acute oral LD (Lethal Dose), spectra data, specific gravity, and boiling point are important. Pesticides were categorized and examined based on ionizability, water solubility, and vapor pressure. Ionizability is the degree to which a pesticide has ionizable functional groups and determines whether it is basic, acidic, amphoteric, or non-ionizable.

4.4. SOLUBILITY

The pH at which half of the chemical is in the ionized form and half is in the nonionized form is known as the pK, or negative log of the dissociation constant, which is a measure of acid or base strength. The term solubility refers to the solubility product (Ksp) in deionized water at a given pH and temperature. The vapor pressure (VP) in millibars of Hg at 25°C is a measure of the tendency of a molecule to enter the vapor state, which is referred to as volatility. Based on sorption tests, soil retention (Koc) is a measure of the pesticide's ability to attach to soil organic carbon. When it was first proposed as a measure of a soil's ability to adsorb non-ionic organic compounds, it was thought that only the soil's organic content was capable of doing so. This is only true, though, for extremely lipophilic substances.

In soils, polar organic molecules are easily bonded to inorganic and organic surfaces. For every given compound, the Koc value is equal to the Kd value divided by the concentration of organic carbon in the soil, or Koc = Kd/Fraction of Organic Carbon. This calculation is done for a variety of soil types. For both ionic and non-ionic compounds, Koc values are frequently provided in pesticide databases, however, the values are rough estimates of the chemical retention by soils. Pesticide longevity is typically measured in terms of the parent molecule half-life under normal conditions in the area in which it is applied. A major problem is the movement of agricultural pollutants through soils and into groundwater. Despite low quantities, prominent pesticide occurrences have been discovered and linked to non-point sources of contamination. Pesticide occurrences and, in certain situations, the absence of pesticides in water cannot be quantitatively predicted based on our understanding of the mechanisms governing pesticide transport and transformations in the vadose zone.

Understanding the possible risk that agricultural chemicals may represent to people and the environment may be the goal of research. Retention, transformation, and transport processes control how a pesticide acts in the environment. Determining the function of these three processes for a particular chemical is important to comprehend the fate of that molecule in the environment. Therefore, estimates of all activities that can potentially affect a pesticide ingredient must be included in assessments of its fate in the environment. The fast transport of water through soil macropores was connected to pesticide migration to groundwater (Köhne et al., 2009). Furthermore, the majority of environmental models neglect the significance of the rhizosphere as a pesticide reaction environment.

The variety of sorption sites that are available will have a significant impact on pesticide transport, because soil is not homogeneous. Therefore, the sorption mechanism and sorption kinetics commonly observed in field conditions would change when uniformly packed soil columns are used. Desorption equilibrium may take longer to attain than adsorption equilibrium. The amount of herbicide in solution may be restricted by this equilibrium. The rate limiting phase in degradation is the rate at which organics desorb from clays into lake water. The crucial mechanism determining the final destiny of chemicals in soil and water is biological degradation, which can take the form of partial breakdown to generate metabolites or total mineralization.

To calculate the environmental fate, an accurate prediction of the degradation behavior is required. Many pesticides can be attacked by soil microorganisms, as has been widely documented in both laboratory and field research. It is challenging to determine the size of the microbial population needed to break down a specific chemical. Researchers have been assessing the degradation potential of laboratory and field soils using plate counts. We might be able to estimate population size accurately with more sophisticated gene probe technology or selective soil biomass classification techniques. In soils improved with carbofuran, the population size has been estimated through the technique of selective labeling. It is obvious that estimating the population in the field is a preferable method. The environment in which bacteria live is simplified in laboratory research, and the four main environmental parameters that affect pesticide degradation are temperature, aeration, pH, and moisture.

4.5. MICROBES

When a microbe tries to use a substance supplied to the soil, the main issue it faces is contact between them. Herbicides adsorbed to soil surfaces and the cells can come into direct contact, or herbicide in solution can migrate to the cells to establish contact. Adsorbed herbicide and microorganisms can only come into direct contact to the extent that the degradation population is large enough (Paszko et al., 2016). The size of the population, the amount of herbicide in solution, and the physical barriers to soil-water diffusion all affect contact by diffusion. Because the experiments produce a good microbe-pesticide interface, they tend to overestimate pesticide degradation that could occur in the field, assuming aeration is sufficient. Researchers demonstrate how an herbicide's rate of degradation dramatically drops as soil moisture falls from 60% water-filled pores (WFP). Herbicide degradation was found to be inversely correlated with the amount of adsorption.

There was a statistically significant relationship between microbial respiration and the concentration of the herbicide in solution (metolachlor). This suggests that where the population is dormant or the concentration in solution is low, the degradation of some herbicides will occur slowly. Herbicide degradation may be influenced by the kind of surface on which they were adsorbed. Montmorillonite clays exhibited a considerable reduction in diquat degradation when compared to kaolinite of the same proportion. 2,4-D was totally shielded from solution and surface-dwelling bacteria once it had adsorbed to soil. 2,4-D was intercellularly metabolized by the bacterium.

The presence of soil organic matter generally regulates the sorption of organic compounds (organics). Herbicides and other pesticides are affected by organic matter in a number of ways. Both the microbial community responsible for degrading herbicides and the adsorption of herbicides into the soil are impacted by organic matter, which includes different unstable non-humic compounds formed by the breakdown of plant wastes. Herbicide availability will be decreased by organic matter propensity to adsorb, although this will not entirely remove the chemical.

Herbicides that bind to soluble soil organic compounds may generate soluble complexes that are not always sorbed. Studies have shown interactions between soluble organic compounds, particularly fulvic and humic acids, and pesticides, including triazines. These kinds of interactions may be crucial in modeling, as they have the potential to drastically alter the fate of pesticides in soil. There have been reports of the microbial biomass having the ability to extract a bound substance from soil. The residues that are discharged may have an impact on plants.

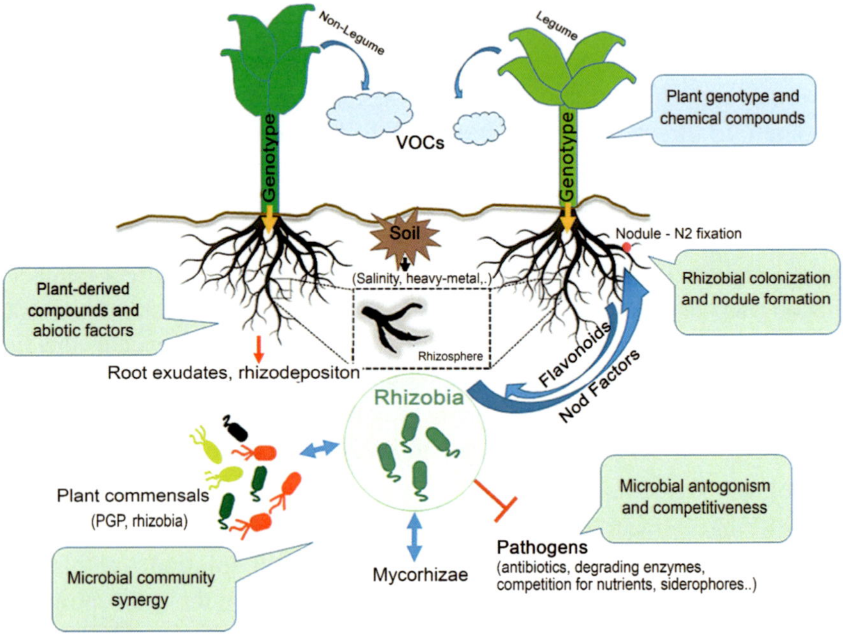

Figure 4.1. *Rhizosphere.*

Source: https://commons.wikimedia.org/wiki/File:Actors_and_interactions_in_the_rhizosphere.jpg, Creative Commons License.

4.6. RHIZOSPHERE

Parathion and diazinon degrade in rhizosphere (Figure 4.1) soils (Jeon et al., 2024). The rates of rhizosphere mineralization of parathion and diazinon were 8% and 10% higher. Certain herbicides reach the roots through the rhizosphere. About 20% of the atrazine was absorbed by corn' most of atrazine is translocated to the shoots where it is detoxified. These metabolites can either reside in the plant waste or return to the rhizosphere. After plant harvest, these metabolites would be added to the soil. Plant metabolites of methabenzthiazuron (MBT) break down more quickly than MBT itself. It is known that the rhizosphere is enhanced by organic components formed by roots, such as exudates, secretions, mucilage, and lysate. In soils, estimates of up to 20% of the material consumed by the plant may be lost as root exudates. The materials deposited by plants in the rhizosphere are high molecular weight chemicals and they interact with pesticides. These organic compounds may enhance the sorption-catalyzed hydrolysis of S-triazines and pesticide adsorption, among other chemical reactions, in the rhizosphere. The herbicide ability to bind to organic materials may prevent it from breaking down. Adsorption, however, has the ability to catalyze non-biological disintegration, which means that it occasionally speeds up the rate of breakdown. The effects of organic matter concentration have been observed to vary and seem to depend on the amount of sorption-catalyzed processes as well as the bioavailability following adsorption. When straw ash was added to the soil, it was observed that atrazine remained more persistent' nevertheless, some researchers found that the degradation of soil modified with charcoal decreased. Recent research indicates a relationship between the content of organic materials and imazaquin adsorption.

4.7. PESTICIDE TRANSPORT

Pesticide transport cannot be explained as a simple function of time or net flow volume since it appears to be flow event-driven through the soil. During less rain, carbofuran concentrations were high. Concentrations rapidly dropped as the flow event continued. The overall response to atrazine and cyanazine was the same, albeit the concentrations were far lower and more unpredictable. With each subsequent drain flow event, the peak concentrations may be low. Pesticide in the soil solution is quickly drained through large pores and into the drain at the beginning of a flow event. Continued water flow through the pores has lower concentrations because desorption is not fast enough to maintain an equilibrium solution concentration in fresh rainwater. Desorption adds chemicals to the water in the large pores when drainage stops, and subsequent rains would result in a fresh flow event with a high initial concentration of pesticides. On the other hand, researchers proposed that drainage from large pores, where chemicals and water have a short residence time, dominates the first phase of the flow event. Consequently, there are significant concentrations in the outflow. Most of the water that is being delivered to the drains when the drain flow slows down comes from smaller pores, where the chemical is in the soil profile for longer periods of time and is consequently found in much lower concentrations. About 14,000 agrochemical companies exist in the United States, which handle, distribute, blend, and administer insecticides and fertilizers. Point source contamination of soil and groundwater has been caused by inadequate containment of pesticide spills and inappropriate disposal of pesticides at some plants. One of the biggest environmental concerns is soil and groundwater contamination brought on by the management, handling, and disposal procedures used in the past and present at agrichemical facilities. Agrichemical facilities frequently have soil and groundwater contaminated by several pesticides at far higher concentrations than those typically found in non-point source contamination. Remedial methods are being developed or currently available to help clean soil and groundwater poisoned by pesticides. Excavated soil may be treated physically, chemically, or biologically, or it may be dug up and disposed of in a landfill authorized to handle hazardous waste (Hull et al., 2005). Incineration, mechanical aeration/extraction, activated carbon adsorption, soil flushing, ion exchange, and membrane separation are examples of physical remediation techniques. Reduction-oxidation, photolysis, and dechlorination are examples of chemical processes.

The goal of land farming is to place pesticide wastes within the plow layer of the soil, where dilution lowers the pesticide concentration sufficiently to facilitate both chemical and aerobic microbial degradation. This is a relatively old technique that has been used to treat municipal wastewater and petroleum sludge. Land farming is now limited to areas where pesticides that are currently approved for agronomic use have contaminated the soil. Authorities that allow land farming mandate that contaminated soils containing non-registered pesticides or Resource Conservation Recovery Act (RICRA) wastes be burned or disposed of in hazardous waste landfills. One of the first states to support laws allowing land farming of pesticide-contaminated soil as a rehabilitation technique was Illinois. An owner of an agrichemical facility may be granted permission by the Illinois Department of Agriculture (IDOA) to remediate a site by applying pesticide-contaminated soil to agricultural land. This authorization was granted by legislation passed in 1990 and in effect until July 1, 1992. When an applicant requests written authorization to apply pesticide-contaminated soils to agricultural land: the IDOA has developed operational control practices or protocols that are mandatory. The Illinois Environmental Protection Agency (IEPA) has instituted an internal procedure to prescribe cleaning objectives and protocols for remedial action at agrichemical facilities. Assuming accountability for remedial measures becomes challenging when there is contamination on a large scale and the source of the contamination is unknown. An accurate evaluation of the degree of soil and groundwater contamination at a site should be the first step in creating a remedial action plan for a particular facility. Establishing protocols will make it easier to conduct site assessments at the land farming and remediation sites. The remedial action plan must include information on the extent of soil contamination, toxicity and occupational exposure to contaminants, the placement and setup of monitoring wells, the amount of soil that needs to be excavated, the excavation process, and the physical and chemical characteristics of the excavated soil. Other elements required for the remedial action plan include the pesticide residue characteristics, concentrations, rates of phytotoxicity, and rates of degradation. In order to avoid reintroducing contaminated soil to the site, a source must be found and a pesticide residue investigation must be done. When choosing backfill materials, it is important to consider the possibility that the site could become contaminated again in the event of another leak. The location of the site and its distance from the source facility, property ownership, any contractual agreements with landowners, an assessment of the vulnerability of surface water and ground water resources, and the possibility of other effects such as leaching, runoff, volatilization, potential for water and wind erosion, etc., should all be considered when evaluating a site that has been chosen to receive pesticide-contaminated soil.

The proposed travel route, the differentiation of the application area and field limits, the identification of municipal, community, or private wells, the soil property characterization i.e., organic matter content, clay content, infiltration rate, hydraulic conductivity, slope, etc., previous pesticide residue levels in the soil, the past and present tillage and cropping history and subsequent agronomic cropping practices, and the timing of the application are additional components.

4.8. REMEDIATION

Protocols for pre-treating the excavated soil should be included in the remedial action plan in case it is required before field application. In order to precisely define the types and amounts of pesticide residues present and calculate application rates, a pesticide residue test of the pretreatment soil must be carried out. It could also be necessary to decontaminate the pretreatment site and the equipment. Various methods have been employed to ascertain the suitable rate of field application for soil polluted by pesticides. One method is to use the maximum concentration of each pesticide residue found in soil pores to determine the pace at which contaminated soil should be applied. Finding the highest possible concentration of pesticide residues in mixed and excavated soil is a useful strategy. Pesticide concentrations in the soil are reduced when the excavated soil is mixed. Smaller application site requirements and higher rates of contaminated soil application are made possible by lower maximum pesticide residue concentrations in mixed soil.

At the application site, pretreatment can also be required. Additional factors to take into account include marking the actual application area and borders, planting and establishing vegetative limits, adjusting soil pH and fertility, choosing primary and secondary tillage treatments, field leveling, and residue management. It is necessary to choose and calibrate application and equipment as well as specify the integration depth for the contaminated soil. As soon as the pesticide is applied, the amounts of residues at the application site should be recorded, and a plan for additional pesticide residue analysis should be created. The best time to apply contaminated soil should be considered when considering the different constituent rates (reference rates) of degradation, the slower rates at which pesticide residues in contaminated soil degrade, and the phytotoxic effects on crops that follow.

Figure 4.2. *Pesticide-contaminated soil.*

Source: https://en.wikipedia.org/wiki/Soil_contamination#/media/File:Soilcontam.JPG, Creative Commons License.

4.9. SOIL CONTAMINATION

The feasibility of applying pesticide-contaminated soil (Figure 4.2) to agricultural land is still being investigated and evaluated (Morillo & Villaverde, 2017). The scope and significance of accurate assessment and characterization of a contaminated site and soil, pretreatment and application considerations, and an assessment of degradation rates of several herbicides following land farming can be demonstrated through case studies from remedial actions. Agricultural pesticides have the ability to permeate the soil profile and enter groundwater, as evidenced by the discovery of DBCP in drinking water wells and the finding of aldicarb residues in drinking water wells on Long Island in 1979. Before these discoveries, scientists were not worried about the potential impact of agricultural chemicals on ground water quality. Furthermore, not much research had been done on the possibility that chemicals used in agriculture could end up in ground water. The discovery of agricultural pesticides in drinking water and other more recent cases led to the start of several computer modeling studies, field research projects, lab tests, and monitoring systems. Determining the amount and scope of agricultural pesticides in drinking water wells as well as creating management strategies to reduce their recurrence in the future were the objectives of research and monitoring projects. The advantage of laboratory experiments over outdoor experiments is that they are typically easier to manage and less expensive. Furthermore, the utilization of radiolabeled materials can streamline sample analysis and enable the identification and quantification of important metabolites. Studies have demonstrated that pesticide behavior in the field may not always be accurately predicted by laboratory results. For instance, assessments of degradation rates in laboratories are often carried out at a fixed temperature, whereas surface soil temperatures in agricultural fields can fluctuate considerably over a 24-hour period. In both the lab and the field, microbial populations could differ. Laboratory tests pertaining to agricultural chemical mobility could be misleading. The impact of soil structure on mobility cannot be replicated in trials using packed soil columns. Some researchers have attempted to maintain the benefits of performing tests in soil columns while simultaneously carrying out similar investigations in the field.

In order to avoid point source contamination, the appropriate use of agricultural chemicals has received more attention throughout the past twenty years. All agricultural pesticides, regardless of their chemical properties, are subject to general management standards to prevent point source residues. Spills close to wells are often the cause of point source residues. Agricultural chemicals should always be handled and transported with caution, and spills should be cleaned. Manufacturers of agricultural chemicals can offer guidance in certain circumstances. The best places to mix agricultural chemicals are on the fields that need to be treated, far from drinking water wells. Any drinking water well should be cleaned before using any equipment. Reinstate from spray tanks must be disposed appropriately, ideally by spraying it on a suitable crop (Damalas et al., 2008). Before being disposed of, empty agricultural chemical containers should be triple rinsed. To stop containers from deteriorating, agricultural chemicals should be stored correctly in a secured area. Farmers should schedule their purchases of agricultural chemicals to prevent prolonged storage. It is also important to have plans in place for handling a storage area (fire). When utilized to put out a fire, water can frequently cause more issues than the fire itself. It is necessary to take precautions to prevent the direct introduction of agricultural pesticides into wells. Management strategies for preventing nonpoint source residues need to be tailored for a particular agricultural chemical, crop, and set of environmental conditions, in contrast to strategies for reducing point source residues. It is possible that procedures designed for one set of conditions won't work for another.

4.10. MANAGEMENT

Optimizing the application, i.e., time, place, and amount of an agricultural chemical is one kind of management approach. Delaying the application of aldicarb to potatoes until after plant emergence reduces the amount of aldicarb residues likely to reach ground water. Local weather patterns must be taken into account while optimizing application scheduling. For instance, because rainfall is highest in the summer in some areas, delaying applications of aldicarb to citrus until the summer increases the possibility that residues will migrate to ground water. If an agricultural chemical is placed close to the soil surface as opposed to farther below, it may degrade more quickly. Reducing the application rate to the minimum required to achieve the intended growth impact or pest control does, however, help to lower residue levels. The flow of agricultural chemicals can also be impacted by irrigation techniques. The quantity of irrigation used should only be sufficient to make up for evapotranspiration losses since rainfall and irrigation water act as the propellants for the downward movement of water and, consequently, agricultural chemicals. Overwatering will cause more water to go lower and therefore more agricultural chemicals. The possible transport of agricultural chemicals may also be impacted by the type of irrigation that is used. For instance, when agricultural chemicals are only present on plant foliage, furrow irrigation may provide less movement than other watering techniques like overhead sprinklers. Sprinkler irrigation reduces mobility of agricultural chemicals in the soil because it distributes water more evenly across the field and makes it easier to apply less water in a single irrigation event.

Prohibiting applications of agricultural chemicals within designated distances of drinking water wells may be a suitable management technique for those compounds that continue to deteriorate upon reaching groundwater. The limitations should take into account the features of the agricultural chemical as well as the current environmental conditions when determining the appropriate distances. In parts of Canada and the United States, regulations mandating buffer zones surrounding shallow wells have been implemented for aldicarb application. Drinking water wells must be placed away from locations where chemicals are loaded, stored, mixed, or disposed. Additionally, the well needs to be at least 15 meters far from any agricultural land. This small buffer zone can lessen the chance that runoff water will carry agricultural chemical residues to the well, protect the well from damage by application equipment, and give any residues in groundwater some time to break down and disperse. Furthermore, wells should not be situated in lower regions where runoff water gathers. Building wells correctly is also crucial. It is important to correctly seal wells at the ground surface to stop any runoff water from traveling down the well casing/wall. Studies show that the top 2 to 5 meters of the saturated zone often contain the majority of agricultural chemical residues that are still degrading in ground water. Well casings should therefore continue uninterrupted for a minimum of 9 meters below the water table.

CHAPTER 5

SUBSURFACE CONTAMINATION

Contents

5.1. INTRODUCTION

Concerns regarding contaminated water supplies first surfaced more than a century ago, as the industrial revolution led to changes in the quality of surface waters. However, groundwater contamination did not become a significant problem until the 1970s since it was widely thought that groundwater was shielded from any kind of pollution. Certain surface waters in the United States have seen improvements in water quality as a result of strict regulation of point-source discharges, which has included the construction of efficient sewage treatment plants. Following the Resource Conservation and Recovery Act or RCRA and the Comprehensive Environmental Response, Compensation, and Liability Act or CERCLA, substantial efforts to clean groundwater started in the 1980s. Early remediation efforts rarely resulted in the anticipated drop in pollution levels. The most common organic pollutants in the subsurface have low solubilities in water and tend to remain as either a separate organic liquid in the subsurface (nonaqueous phase liquid, or NAPL) as in the case of chlorinated solvents or a separate solid phase as in the case of chemical explosives that have precipitated in the subsurface. The heterogeneities in the subsurface, sorption of contaminants onto solid organic matter, and contaminant diffusion into low-permeability zones combine to make treatment considerably less efficient than initially thought. Denser than organic liquids are called dense nonaqueous phase liquids (DNAPLs) (Yang et al., 2021). It became apparent in the late 1980s that a site would be very challenging to clean if DNAPLs were present.

5.2. CONTAMINATION

The minute quantity of contamination that dissolves will contaminate groundwater that flows through the volume of subsurface holding the contaminant, known as the source zone. This implies that until the contaminating source is eliminated or physically isolated from flowing groundwater, groundwater returning to pollution free state will not be accomplished. It is difficult to locate contamination sources within the subsurface due to a lack of efficient characterization methods and the tendency of DNAPLs to have a geographically confined but extremely heterogeneous distribution. Furthermore, even though a few new technologies have been created to clean source zones, it is difficult to assess these technologies, which makes it difficult to forecast how effective they will be in large-scale applications. Currently, pump-and-treat systems are in place at the majority of DNAPL locations to contain dissolved phase plumes and reduce public hazard. When pollution is identified, there is frequent public pressure to take remedial action. The fact that remediation technologies for DNAPL sites are expensive, meaning that there haven't been many large-scale remediation attempts puts pressure on sites to be cleaned up. Whether it is worth the expense to undertake remediation at DNAPL sites depends on the objectives of the remediation project, on what can be achieved, and on the competing needs of other critical sites. As the anticipated cost of cleaning up the contaminated groundwater ranges from a few to several hundred billion dollars, it is crucial to choose the places where remediation activities can have the impact.

5.3. REGULATIONS

Several government organizations have calculated the long-term costs of keeping pump-and-treat systems in place despite their inefficiency' these estimates range from the hundreds of millions of dollars to billions of dollars. A variety of explicit regulations published by federal and state regulatory agencies led to the acceptance of more containment as a response to the mounting expenses associated with cleaning up polluted sites and the growing realization of the limitations of technology. There have been transformed efforts to try to shorten the time needed for remedy operation and monitoring by trying to remove a significant portion of contaminants at many hazardous waste sites with more aggressive source remediation technologies. Both standard and novel ex situ and in situ technologies can be used in source cleanup. Despite a general trend toward decreased private investment in cutting-edge technology during the 1990s, the use of more aggressive source remediation approaches has expanded significantly over the years.

The more problematic organic compounds at some installations are those that are difficult to break down, like vinyl chloride, dichloroethene and dichloroethane, which can all be found in subsurface DNAPLs. These compounds include the chlorinated solvents perchloroethene, trichloroethene, and trichloroethane. Military equipment was frequently cleaned and degreased with chlorinated organic solvents, which were usually disposed of in drums or at the surface. Coal tar, polycyclic aromatic hydrocarbons (PAHs), and polychlorinated biphenyls (PCBs) are other commonly reported hard-to-treat organic substances at many locations. Prior to their use being prohibited, PCB mixtures were utilized as dielectric fluids in electrical transformers and capacitors.

Petroleum products contain PAHs, whereas coal tar and creosote, which were once widely used to treat wood, are combinations of hundreds of different molecules, including nitrogen-heterocyclic compounds, phenols, and naphthalene and other PAHs. Along with paints, perchlorate, bis (2-ethylhexyl) phthalate, nitrates, heavy metals (lead), and pesticides and herbicides, these substances are also commonly observed at some locations. The Army has the greatest number of locations impacted by chemical explosives.

5.4. PROCESSES

Many different geological processes result in a multitude of differences in subsurface environments. Eolian (wind-blown) sands, alluvial fans, lacustrine clays are examples of common sedimentary systems. Common rock systems include granites, limestone, dolomite, sandstone, shale, interbedded sandstone and shale, and metamorphic systems of crystalline rock (Bustillo Revuelta, 2024). It is difficult to establish conclusions about the behavior of source zones, the effectiveness of remedial treatments, and the attainable results because of this variability. For instance, the methods needed to define alluvium differ significantly from those needed to characterize rock, and the movement of remedial fluids such as surfactants or groundwater in beach sand differs significantly from those in karst systems. Under both natural and artificial conditions, these characteristics regulate the processes by which pollutants are stored and released from source zones. Source zones can occur in more than one hydrogeologic setting, such as alluvium covering fractured crystalline rock, or they can occur within a single hydrogeologic setting, such as a sand dune deposit.

The qualities of the organic fluid, such as density, viscosity, and interfacial tension with the pore water, as well as the features such as surface characteristics and texture, control bulk NAPL migration within a formation. A NAPL will initially come into contact with the unsaturated zone, which has natural pore water and air, as it descends into the subsurface. Because gravity is the main force for NAPL migration in this zone, which involves the NAPL displacing air, the migration will often be vertically downward. As the NAPL travels in the unsaturated zone, it may dissolve into the pore water or volatilize into the pore air, depending on its solubility and volatility. The saturated zone, where water entirely fills the pore space, will eventually reach NAPL as it migrates downstream. DNAPL tendency to migrate vertically is due to its density' this causes the groundwater to be displaced until a less permeable stratum is reached, the volume of DNAPL is reduced, or a sufficient vertical hydraulic gradient is satisfied. LNAPL will typically migrate in the direction of natural groundwater flow and spread preferentially within the capillary fringe zone. It should be noted that chlorinated solvents have a tendency for vertical flow since they have densities of 1.2 g/cm^3 or more (Kueper et al., 2014). However, the densities of coal tar, and other PAH-based DNAPLs are typically significantly closer to that of water, meaning that they might not experience pressure for vertical migration. NAPL viscosity affects its migration rate, whereas its density determines its tendency to migrate vertically inside the saturated groundwater zone. Lower viscosity fluids typically move more quickly because there is less resistance to flow. Many DNAPLs, especially chlorinated solvents, have viscosities lower than that of water. These liquids will therefore often flow easily. However, with similar hydraulic gradients, coal tars often have a higher viscosity and travel more slowly.

5.5. CAPILLARITY

Capillarity is a characteristic that affects DNAPL mobility in the subsurface. The interfacial forces that exist between phases physically appear as capillarity. Interfacial tension and wettability, two interfacial parameters, as well as the size of the pores, regulate capillarity. The energy per unit interfacial/cross section area needed to form a new surface is known as interfacial tension, which is a characteristic of the aqueous DNAPL. For pure organic liquids, DNAPL–water interfacial tensions usually lie between 20 and 50 dynes/cm. However, co-contaminants or additives in the DNAPL phase, such as organic acids, bases, and surfactants, as well as dissolved pore water constituents, including naturally occurring humic compounds, can have a substantial impact on interfacial tension. These substances have the ability to operate as surface-active agents and significantly lower interfacial tension. A decrease in interfacial tension tends to reduce the effort needed for a DNAPL to displace water from a saturated pore as well as the spread of the DNAPL transverse to, or perpendicular to, its primary direction of migration. The term wettability describes a fluid's propensity to spread out or stick to a solid surface when another immiscible fluid is present. It regulates how fluids are distributed inside the pores. The liquid that coats the solid in water/NAPL/solid systems and has a greater affinity for the solid surface is called the wetting phase, and the other liquid is referred to as the non-wetting phase. The angle formed by the liquid-liquid interface with the solid surface is known as the contact angle, which is wettability metric. Water is drawn to the surface of many natural minerals, such as carbonates and quartz, more strongly than it is to the constituents of typical DNAPL. Water typically serves as the wetting phase in these types of media, dispersing throughout the solid surfaces, and cracks.

The surface is considered to be strongly NAPL wetting when the contact angle becomes closer to 180°. If the contact angle of a surface falls between roughly 70 ° and 120 °, it is referred to as intermediate-wet. In the petroleum industry, the state of mixed wettability, where the larger pores are organic-wetting and the smaller pores are water-wetting has long been acknowledged. Media with surfaces that vary in wettability are typically referred to as having slight wettability. Natural changes or the interaction of released NAPLs with solids can result in the simultaneous existence of water-, intermediate-, and organic-wetting conditions in the subsurface. For instance, contact with NAPL mixtures that contain components that are surface-active can make a porous medium wet, ranging from intermediate to organic. NAPL migration and persistence in natural environments may be impacted by such modifications. While capillary forces for basic pore structures (cylindrical pores) are well understood, accurate interface position predictions are challenging due to the diverse pore structure of natural porous media. Therefore, the capillary behavior of a given medium is usually described by a macroscopic relationship known as a capillary pressure–saturation curve, which relates the liquid pressure difference to the wetting fluid's degree of saturation. The entry pressure is the necessary pressure differential that must be reached in order for any wetting fluid to be displaced.

5.6. DNAPL

DNAPL frequently travels in a very unpredictable manner, producing a source zone with thin horizontal lenses and vertical channels that are related to one another. This will be especially true for Type IV or V media with large horizontal cracks, or for a Type III geologic environment with continuous fine-textured horizontal layers of high permeability variation. Even if the hydraulic gradient is pointing in a different direction, the DNAPL will likewise tend to flow down-dip when the impermeable or fine textured layers are dipping. A DNAPL may be displaced far from its point of entry by this combination of down-dip migration and horizontal spreading until a more permeable channel is discovered. A portion of the DNAPL may be trapped by variations in the topology of fine-grained layers, resulting in isolated pools. Capillary forces cause tiny globules of this organic phase to be held inside the pores while the DNAPL migrates across the subsurface. The volumetric ratio of the entrapped organic phase to the total pore volume is known as the residual saturation, and it is commonly used to quantify this entrapped NAPL. When capillary forces are great enough to oppose the forces of gravity and flowing water, entrapment happens. Therefore, pore geometry, organic phase characteristics (interfacial tension, viscosity, and density), flow velocity, and porous media wettability all affect residual saturations. The entrapped residual is also dependent on the maximum DNAPL saturation obtained prior to drainage of the organics. Furthermore, a volume of DNAPL will often move deeper and farther into a formation if it is released more slowly because larger amounts of an organic molecule are entrapped at higher release rates. As a result, a DNAPL spill migratory patterns are typically influenced by release rate. It should be mentioned that a break in the capillary barrier that supports the pool or a subsequent alteration in the groundwater flow field have the potential to mobilize stored organic matter. However, residual saturations are projected to be basically independent of velocity at groundwater velocities typically seen under natural or pump-and-treat conditions.

DNAPLs local maximum residual saturations, as determined by field- and lab-based tests, usually fall between 10 and 35% in saturated, unconsolidated media and reach up to 50% in low-permeability materials. However, average residual saturations recorded at common DNAPL field sites are typically significantly smaller, typically between 0.1 and 1.0% of the affected pore volume (Dearden et al., 2013). The challenge of predicting and evaluating the storage and elution of contaminants from low-permeability zones is intricate because of the vast variations in groundwater flow velocities seen in various interbeds of natural sediments, such as sand and silt. The sorption process is usually viewed as an equilibrium process when groundwater flow is rather slow, as is often the case under natural flow conditions. This means that these solid-water interactions occur immediately as the water passes through the solids. Desorption may also have a constraining effect on the biological and chemical pollutant transformation in the aqueous phase, particularly in cases where the transformations occur quickly.

Because of the toxicity of high contaminant concentrations linked to the presence of NAPL, it was previously thought that biotransformation processes could not occur close to a chlorinated solvent source. However, a number of investigations have reported microbial activity at concentrations at or near the aqueous solubility of PCE. Although enhanced dissolution has not yet been thoroughly documented in the field, recent studies have suggested that microbial activity can enhance DNAPL dissolution by a factor of 5 or more in the laboratory by increasing local concentration gradients and enhancing aqueous solubility. When DNAPLs are resistant to normal biodegradation and are only moderately soluble in water: plumes form downstream of the source material. For instance, chlorinated solvents sorb weakly on aquifer materials and are not readily broken down by oxidative microbial processes' instead:, they are more likely to be biodegraded, at least in part, through reductive dechlorination or co-metabolic oxidative reactions. Chlorinated solvents can therefore create large plumes that extend downstream from the source. The amount to which distinct biogeochemical processes have taken place can determine how different the chemical composition and concentration range of the pollutant in the plume is from that of the source zone. Plumes will typically take on spatial characteristics closer to the source zone, which will resemble the uneven saturation distributions. If the conditions around groundwater flow are unstable, as they may be due to seasonal variations in infiltration, the form and size of a groundwater plume may alter over time. In these circumstances, the source zone, which is losing mass slowly enough to appear constant over the course of the observation period, feeds the plume. Mass is lost or, more precisely, attenuated to below detection limits on the periphery of the plume due to transformation processes like biodegradation or groundwater dilution mechanisms like advection and dispersion. It should be mentioned that in many hydrogeologic environments, pollutants from the plume would typically sorb or diffuse onto aquifer solids, followed by reverse diffusion. This behavior may contribute to the long-term contamination of groundwater. Not every plume of contaminants in groundwater indicates the existence of a source zone.

5.7. TNT

The most common explosive used in military ordnance is TNT. During World War II, the United States' production of TNT reached a height of 65 tons per day and is limited to military arsenals. TNT is one of the most stable explosives when purified, and it works well after storage for a long time. Sequential nitration of toluene is the first step in the commercial manufacturing of TNT. Toluene is heated and sulfuric and nitric acids are added to make mononitrotoluene, often known as MNT. Utilizing nitric acid, this is transformed into dinitrotoluene. Bi-oil was cooked with sulfuric and nitric acids reinforced with oleum 2 before being converted to TNT. The batch procedure generates a mixture of chemicals at each stage. The pre-product TNT is melted, and the other less desirable isomers are separated from 2,4,6-TNT by washing it with soda ash solution and sodium sulfite. Each process's waste acids are directed to the batch stage before them. A network of flumes transports waste water/effluent from the last purification stages to a treatment area, which is defined as explosives-contaminated water. Water-filled tanks are used in both batch and continuous TNT manufacture at each production house to halt the process in the event that an uncontrollable reaction starts. Significant sources of explosives contamination below the surface could include spills, material discharges to the tanks, and leaks in the transportation of materials between production houses and from storage in holding tanks. MNTs are substantially less dense than chlorinated solvents, with a specific gravity that varies according to the isomer and ranges from 1.155 to 1.160 g/cm^3. The high sulfuric acid content, which modifies the material density and the component solubility, further complicates the physical-chemical characteristics of the TNT process discharge materials. The production of DNT often resembles that of TNT. As a result, the same variables that govern the physical-chemical characteristics of any effluent material from DNT production also govern TNT production. Different DNT isomers are produced via nitration of MNT isomers. For instance, whereas nitration of p-nitrotoluene yields just 2,4-DNT, nitration of o-nitrotoluene yields both 2,4- and 2,6-DNT. In the process of producing 2,4,6-TNT, nitration of m-nitrotoluene yields 3,4-DNT, 2,3-DNT, and 3,6-DNT, all of which are unwanted byproducts. RDX stands for Research Demolition explosive/Royal Demolition explosive." Synonyms for RDX include cyclotrimethylenetrinitramine, hexahydro-1,3,5-trinitro-sym-triazine, and cyclonite. RDX is a white, crystalline substance with a 37.84% nitrogen concentration. RDX is typically combined with oils, waxes, or other explosives

5.8. MICROBIAL DEGRADATION

Under both aerobic and anaerobic conditions, naturally occurring microorganisms and soil minerals quickly convert TNT to mono-and then di-aminonitrotoluenes. The majority of TNT biodegradation seems to be co-metabolic. Because the aromatic ring becomes unstable and the remaining nitro groups become less electrophilic, reduction rates drop as each nitro group is reduced. Some of the aminonitrotoluenes may still engage in interactions with soil organic matter and form covalent bonds that lead to a multi-step humification process. Under aerobic conditions, certain bacteria can biodegrade 2,4-DNT and the 2,6-DNT production impurity, using these molecules as sources of carbon, nitrogen, and energy. The most potent environmental interactions with RDX occur when reducing circumstances are met, which subsequently reduce the nitro groups to RDX products. Ring cleavage then results in the production of a range of short-chain compounds. The destiny of HMX in the subsurface has not been thoroughly studied, although it is thought that it will undergo ring cleavage and a sequential reduction of the nitro groups, just as RDX. The environment prefers to lose TNT>DNT>RDX>HMX due to natural reduction mechanisms. Continuous elimination reactions are made possible by TNT sorption and aerobic decomposition. Both HMX and RDX absorb poorly to soils and need conditions that are significantly reduced in order to undergo natural elimination reactions. Because of this, together with its toxicity, RDX has emerged as one of the more difficult organic explosives to deal with.

Excavation, containment, and pump-and-treat are typical methods for dealing with DNAPL contamination (Khaitan et al., 2006). One method to extract as much of the mobile DNAPL as feasible is multiphase extraction. The remaining approaches are classified as either extraction, transformation, or both, and they target remnant DNAPL. While transformation technologies aim to change the form of the DNAPL in situ, extraction technologies try to increase the rate at which the DNAPL can be collected from the subsurface. Many technologies are dual-purpose. The traditional methods of pump-and-treat and excavation are important in the management of source regions at hazardous waste sites. In a way, confinement and excavation represent the extremes of source clean-up techniques. Using large construction machinery capable of extracting raw materials and putting them into shipping containers, excavation is done. After that, the containers are transported to a suitable location, which can include specific on-site locations, for handling or disposal. Backfilling the excavation is necessary, and in order to prevent cross-contamination, clean backfill material must be readily available and placed properly and safely. A substantial amount of physical access to the source location is necessary for each of these tasks. Knowing the location, depth, and overall distribution of source materials is crucial for the success of excavation, which calls for a thorough source characterization before work even starts. The materials encountered during the excavation are classified as contaminated or uncontaminated using the same characterization tools, which are also used later to confirm that all of the source material has been removed. Basic geotechnical knowledge is crucial for forecasting the success of excavation, in addition to details regarding the dimensions and form of the source zone.

5.9. EXCAVATION

Since bedrock is difficult to extract, one should ascertain whether it is present. Because groundwater is contaminated from contact with the source material and needs to be cleaned, excavation below the water table is challenging. The term 'running sands' refers to the tendency of saturated sandy soils to liquefy during excavation, which can significantly increase the complexity of the process. In certain conditions, sheet pilings or dewatering systems around the source materials are required to control water flow and stabilize the base and side walls of the excavation.

Figure 5.1. *Excavation.*

Source: https://en.wikipedia.org/wiki/File:Hdpe_pipe_installation.jpg, Creative Commons License.

Having sufficient physical access to the source zone is crucial for the successful excavation of a source. Should adjacent buildings or foundations impede physical access, total demolition might not be feasible without causing significant damage to neighboring structures. Whenever an excavation is scheduled in close proximity to a building or foundation, it is very crucial to conduct a thorough examination prior to the excavation process. Additionally, the construction equipment tends to slip in risky ways on steep slopes with a thin coating of contaminated soil, making excavation challenging. Excavation is a more viable restorative approach in some hydrogeologic environments. Sites with thin confining layers, low soil permeabilities, a volume of source materials less than 5,000 m^3, and contaminants that don't require complicated treatment or disposal are the most favorable for excavation in terms of both effectiveness and cost. The primary method for treating soils heavily polluted with explosives is now composting, which requires the excavated material to be combined with clean soil as a pretreatment for very high explosives concentrations.

Excavation (Figure 5.1) has the main benefit of rapidly removing source materials from the groundwater system. As soon as excavation is completed, the flow of contaminants out of the source area is stopped. Excavation is frequently chosen by stakeholders and potentially responsible parties (PRPs) over in situ treatments, even if it may be less expensive (Lin et al., 1996). The necessity for a location that can hold the material that is excavated, the risks associated with using heavy machinery, the possibility of worker exposure to volatile organic compound (VOC) releases, and the inability to forecast source area volumes are just a few of the drawbacks of excavation. In fact, due to inaccurate initial source characterization, operations frequently remove up to twice the volume of source material projected prior to the excavation.

When water tables are dropped for excavation, it is possible that DNAPL will be remobilized and will flow into the excavation, creating a hazard (workers). Cost can be a drawback if the excavated volume is considerable or if the source materials excavated are subject to land disposal restrictions that lead to significant ex situ treatment costs. A planned and executed excavation carried out in an appropriate hydrogeologic setting should entirely remove all mass in a source zone. In these cases, mass flux reduction, concentration decrease, and reduction of source migration potential will also be satisfied. Excavation creates no change in contaminant toxicity since the contaminants are taken off site for treatment or disposal. The purpose of a containment strategy is to limit hazards by greatly decreasing contaminant migration via containment of the source such that there can be no direct path of exposure to the source. Physical containment is accomplished by building impermeable barriers on every side of the source zone with typical heavy construction procedures and equipment. Thus, a common containment remedy consists of a very low-permeability vertical barrier surrounding the source on all sides, a clay aquitard below the source, and a low-permeability cover on top. Vertical barriers can be created utilizing bentonite slurries, slurries combined with polymer sheets, sheet pilings with sealed joints, pressurized injection methods, or cryogenic systems that freeze the soil. Top barriers/covers are utilized to reduce entry of rain water and subsequent dissolution of pollutants. Most top barriers are layered systems that include polymer sheeting and drainage layers. Typical operating practice for a containment system is to keep groundwater levels inside the container low relative to the adjacent aquifer by operating a small pump-and-treat system that withdraws groundwater from inside the system. By doing this, an inward gradient in groundwater is created, which helps prevent toxins from migrating outward. Top barriers play a major role in preserving an inward gradient and bringing down the cost of pumping and treating.

5.10. SAFETY ISSUES

The primary containment safety issues are related to using the large machinery required for construction. Maintaining the effectiveness of a containment system once it is established is crucial to preventing potential health issues or other difficulties in the surrounding areas. This calls for thorough and ongoing inspections while the remediation structure is being built, as well as long-term follow-up/supervision. Although the area can no longer be used for any intrusive activities, newer construction technologies like jetting are reducing land-use restrictions around contained systems, and alternative land uses like parks and golf courses built on top of containment systems are becoming more common. In order to prevent further exposures, a system to regulate gas migration should be constructed if the source materials have the capacity to produce gases. A well-designed containment system virtually reduces the transfer of contaminants to other environmental compartments, hence preventing exposures, both direct and indirect. Containment systems offer a very high degree of mass flux reduction and a very high reduction of source migration potential. Nevertheless, to ensure that pollutants do not migrate, containment system monitoring is crucial. Except when intentionally utilized along with treatment technologies, containment systems have no effect on the mass, concentration, or toxicity of the source zone. Containment systems and in situ treatment can work together since most in situ technologies can clean-up a source while operating inside a contained zone.

One of the most popular techniques for preventing contaminants from leaving DNAPL source zones is hydraulic confinement. Contaminated groundwater from source zones can be extracted and treated ex situ using extraction wells' this process is known as pump-and-treat (Ciampi et al., 2023).

Injection wells can be used along with extraction wells to hydraulically separate contamination source zones, hence lowering the expenses associated with ex situ treatment. It is widely acknowledged that the limited solubilities of the majority of contaminants of concern and the limitations in mass transfer to the aqueous phase make hydraulic containment generally ineffective for source remediation. Selecting the best well sites and pumping rates necessitates a deep understanding of the hydrogeology of the site in order to design hydraulic containment systems effectively. Inadequate hydrogeologic characterization can result in systems that pump too much groundwater or fail to achieve total containment, which raises the cost of pumping and treating. Therefore, the design of an ideal hydraulic containment system will be more difficult in a complicated hydrogeologic situation. Other than those related to well drilling, hydraulic containment has no depth restrictions' nonetheless, as well depth increases, prices should also raise. Because large pumping rates from closely spaced wells may be necessary in systems with high hydraulic conductivities, hydraulic containment may be challenging to achieve. Because large capture zone sizes need steep gradients, achieving efficient hydraulic containment may also be challenging in low-permeability formations such as clays and silts. Because lower-permeability zones cause a lack of hydraulic connections, effective hydraulic containment in highly diverse systems is constrained. This could be especially difficult for karst and fractured systems, where it can be challenging to determine connectedness. Hydraulic containment primarily addresses health, safety, and environmental issues with regard to the removal and disposal of sub-surface extracted toxins. It is imperative to take precautions to prevent exposure to extracted polluted groundwater, especially in the event of contaminant vapors gassing. Catalytic oxidation, biological treatment, and activated carbon are common ex-situ treatment procedures. The extra procedures required for the final disposal of contaminants that are transferred to another media, such as activated carbon treatment, may pose health and safety issues.

5.11. VADOSE ZONE

The vadose zone is typically where multiphase extraction wells can be built (Gabr et al., 2013). In order to eliminate volatile organic vapors from the soil gas, vapor flow is created by the applied vacuum and travels through the vadose zone to the multiphase extraction wells. Water extraction lowers the water table, which increases the subsurface vulnerability. Groundwater extraction also clears the subsoil of dissolved pollutants. Without significantly lowering the water table, applying a high vacuum to the extraction well increases pressure gradients surrounding the well and improves groundwater flow to it. When there is a large drop in the water table, spreading of LNAPLs in the soil surrounding the well can be lessened if LNAPL is present at the site. Determining the zones of impact of wells for specific vacuum levels, gas and liquid extraction rates, and the ideal well spacing are all necessary for designing a multiphase extraction system. Hydraulic models for gas and water flow can be used for preliminary design, although pilot testing is recommended. Pumps, gas-liquid separators, and trains for gas and liquid treatment are among the aboveground equipment requirements. There are several unique designs for multiphase extraction that have been created. These designs usually differ in the specifics of how the fluid is extracted from the wells.

5.12. SUPERSOLUBILIZATION

Supersolubilization is an alternative technique that exists between the two extremes of solubilization and mobilization, which are commonly regarded to be the fundamental mechanisms behind surfactant/co-solvent flows. Mobilization and vertical migration are prevented by maximizing solubility enhancement while preserving enough interfacial tension.

The optimal strategy for a particular site is often determined by site-specific factors. The compatibility of the additive with the subsurface environment, which includes the porous media, groundwater, and the NAPL itself, should be taken into account while constructing a surfactant system. Ignoring these interactions could lead to an excessive amount of additive loss (phase separation or partitioning into the NAPL or surfactant sorption or precipitation or in combinations), which would make the system extremely inefficient. The viscosity and density of the flushing solutions, both before and after contacting the NAPL, the effects of temperature and salinity/hardness on the system, the biodegradability of the additive and its metabolites, and the potential effects of the additive, both in the flushing zone and in other regions that may be impacted, are additional factors that must be taken into account when designing these systems.

Highly heterogeneous systems may call for specific design elements such as the usage of foam or polymers or unusual hydraulic schemes like vertical circulation wells. After remediation, the site is flooded with water, regardless of the method, to remove related pollutants and surfactant/co-solvent. Hazardous chemicals including ozone, peroxide, permanganate, and persulfate need to be handled carefully. Application of Fenton's reagent or ozone can produce a large amount of gas and excessive heat. In particular, since ozone is a gas, more care must be used. Acidity is produced by the oxidation of pollutants and organic materials in the soil, which can lower groundwater pH if adequate buffering capacity is not present naturally or is not provided. Additionally, exchangeable and redox-sensitive sorbed metal ions may be mobilized.

5.13. OXIDATION

Many of these oxidation reactions might be harmful to the environment. Oxidants might not be the best option if natural attenuation is required as a polishing step following the source remediation phase because they may wipe out native microbial populations, especially redox-sensitive anaerobic microbial communities linked to chlorinated solvent biodegradation. Fenton's reagent application was used to decrease soil and groundwater microbial populations, especially methanotrophs (Merino et al., 2021). Minimal research has been done on how oxidation technologies affect naturally occurring subsurface microbial activity. The production of colloidal materials may result in permeability reductions due to in situ chemical oxidation. Manganese dioxide is precipitated as a result of permanganate reactivity with organic matter. This can decrease soil permeability, persist as a long-term source of manganese in the soil, and create problems in some sensitive ecosystems. After subsurface conditions are established, injection well sites are identified, and oxidant requirements are approximated, the installation and operation of in situ chemical oxidation technologies are rather simple. To assess the possible efficacy of oxidants for the soils and contaminants to be treated, pilot-scale testing is advised. Monitoring of groundwater geochemistry i.e., pH, redox, dissolved metals, oxidant concentrations, reaction products like chloride, and temperature should all be taken into consideration while evaluating the efficacy of chemical oxidation. Post-oxidation monitoring should also be carried out to assess potential rebound of microbial populations, release of metals, dissipation of oxidants, and rebound of pollutant concentrations.

Chemical oxidation systems typically fail due to inadequate delineation of pollutant distribution in the subsurface or inefficient delivery of oxidants resulting from subsurface heterogeneities. The efficiency of destroying contaminants can be severely hampered in heterogeneous soils due to the movement of oxidants into low-permeability zones where contaminants are present. The use of chemical oxidants in fractured rock and clay has received less attention' however, because of the diffusion-limited mass transfer rates of oxidants into the clay and rock matrices, especially in the case of the unstable ozone and peroxide oxidants, these techniques are not expected to be very effective in these environments. The efficiency of in situ chemical oxidation might be further decreased by altering subsurface permeability as a result of gas or colloidal material production, or, in the case of permanganate, from precipitation of manganese dioxide. The design of in situ chemical oxidation systems necessitates the prediction of target contamination removal rates in addition to the selection of injection well spacing and injection rates. Since peroxide and permanganate are injected in aqueous solution, injection systems for these oxidants can be built using standard groundwater models. Vapor flow models can be used to inject ozone into the vadose zone (Kim & Choi, 2002). However, there are no reliable models available to anticipate the movement of sparged gases accurately when ozone sparges into the saturated zone. It is necessary to simulate the distribution of pollutants, the flow of oxidants that are injected, as well as the interactions between oxidants and contaminants, in order to predict the rates of contaminant oxidation.

CHAPTER 6

ADVANCES IN GROUNDWATER REMEDIATION

Contents

6.1. INTRODUCTION

There are two dynamic formulations that illustrate the application of genetic algorithms to the dynamic optimal design of pump-and-treat groundwater remediation systems. The variables in the first formulation represent the durations of the management periods, where it can be assumed that the contaminant sorption was in equilibrium. The second formulation can use mass-transfer-limited sorption and assumes an intermittent pumping strategy to eliminate a pollutant. Even though the genetic algorithm was able to tackle these dynamic challenges, the overall remediation costs were only reduced by a minute amount. The mass transfer-limited pulsed pumping can save up to 10% compared to continuous pumping, while the flexible management periods save up to 3% compared to fixed periods. Nevertheless, the operating cost savings were more considerable. Even a minute proportion of operational cost reductions could add up given the high price of remediation. For example, allowing for varied management periods achieved a cost reduction of almost $26,000, which correlated with the 3% savings with flexible management periods. Especially for sites with limited mass transfer, dynamic pumping that is gradually adjusted to the particular site conditions can be a way to increase the remediation cost-effectiveness. In addition to offering policies that are better than those that have been put into practice, optimum design can also be an effective method for exploring other scenarios or variables. An optimal design can show if a project is feasible or not. Many optimization techniques have been used to design groundwater remediation systems. Recently, the best groundwater remediation design has also been achieved through the use of genetic algorithms, or GAs. A random search algorithm employs probabilistic search criteria. This probabilistic search method was found to be successful in identifying optimal or nearly optimal solutions in works that applied GAs to optimal groundwater remediation. GAs have been used to address multi-objective groundwater issues, including confinement and groundwater monitoring (Luo et al., 2016). Using GAs in groundwater remediation has several benefits. Given the variety of field conditions and the vast array of remediation solutions, flexibility is crucial in remediation design. The best steady-state strategies for groundwater reclamation have been proven to be more expensive. Dynamic groundwater remedial design has been the subject of theoretical approaches which are specifically designed to handle differential dynamic systems.

6.2. PULSE PUMPING

The use of pulsed pumping, which involves regularly turning on and off pumps, has been proposed as a way to lower groundwater remediation costs. This could be helpful for sites with limited mass transfer, like those that have been contaminated for a long time, as well as for in situ bioremediation systems. The rate at which a contaminant transitions from the sorbed phase to the aqueous phase may be limited by slow mass transfer. Therefore, even if a pump-and-treat system might remove a large amount of contamination at first, since slow mass transfer limits the concentration in the aqueous phase, the extracted water concentration would start to decrease. The outcome of continual pumping is the extraction of clean water. Because resident time can lower operating costs by allowing the concentration gradient between the aqueous and sorbed phases to push the sorbed contaminant into the aqueous phase, pulsed pumping may be a cost-effective method of pumping. In contrast to continuous pumping, a lot of research on pulsed remediation techniques has indicated less improvement in overall mass removal efficiency. The cost advantages of pulsed pumping over continuous pumping are investigated using GAs. The design of groundwater plume control systems in unconfined aquifers was demonstrated using a combined simulation and optimization model on two- and three-dimensional aquifers. Mathematical difficulties arise in unconfined aquifer simulation that do not arise under presumed confined conditions. Excessive extraction from a pumping well in an unconfined aquifer might cause the well to dry out and stop pumping. When designing containment systems in unconfined aquifers using simulation-optimization models, it is important to take into account both the nonlinearity and the possibility for dewatering. Reducing the overall pumping of the containment system, while controlling plume migration is the aim of the groundwater management plan.

6.3. CONTAMINATION PLUMES

Groundwater contamination plumes can pose serious risks to both human and ecological health if they are allowed to spread through the subsurface and reach surface water bodies or wells. Many solutions have been developed in the last few decades to manage and clean up subsurface contamination. The pump-and-treat approach, which uses a system of injection and extraction wells to manage the subsurface flow regime such that contaminants are trapped and removed by extraction wells, is one of the extensively used remedial treatments. Pump-and-treat design entails a series of choices that ultimately impact the project cost and success, including determining the location and size of pumping valves to be placed on the aquifer system. However, it can be very challenging to employ trial-and-error to find a set of constraints that meet remedial goals and are also cost-effective at many groundwater contaminated sites due to the hydrogeologic complications. Combined optimization-simulation models are effective tools that enable the engineer to work on several project goals at once, including finding the least expensive solution that meets all requirements. An alternative to trial-and-error design is offered by optimization techniques, which call for the design problem to be expressed mathematically as a set of project objectives and constraints. It is possible to conduct a systematic search within the feasible engineering designs for the optimal solution that best satisfies the specified management objective, such as cost minimization, by incorporating mathematical optimization tools into the engineering design process. Sometimes competing constraints make a problem unfeasible' in these situations, optimization models can definitively identify the solution, while a trial-and-error technique may find it nearly hard to distinguish.

6.4. OPTIMIZATION MODELS

The three main groups of optimization models are concentration control models, advective control models, and hydraulic control models, which are used for groundwater plume containment design. The groundwater flow system must be simulated in order to use any of the three groups of optimization models. Advective control models add a particle tracking technique to simulate advective transport, concentration control models necessitate an advective-dispersive contaminant transport model based on concentration, and hydraulic control models are built using just the flow simulator. The pump-and-treat system's well locations and pumping rates serve as the design variables in each scenario. An optimization method is then utilized to determine which design best satisfies a given management goal. There are a number of reviews of optimization-simulation models for managing groundwater quality that are accessible. By restricting hydraulic gradients or velocity vectors at certain points within the model domain, hydraulic control formulations indirectly limit plume migration (Mulligan & Ahlfeld, 1999). In order to stop plume migration beyond some locations, the constraint values are selected to drive the flow direction at the edge of the plume interior and toward extraction wells. Hydraulic control solutions should be verified using post-optimization simulation to make sure that the control system is actually collecting the plume, as hydraulic restrictions serve as substitutes for the actual design purpose of doing so. Concentration control optimization models usually constrain the design stresses to satisfy specific concentration standards within the aquifer at specific times. Due to their nonlinear characteristics, these models are challenging to solve, and a wide range of optimization techniques have been studied in an attempt to find a solution that can withstand these mathematical difficulties. The optimization formulation searches for a pumping system that can either completely enclose the contaminated plume in a capture zone or successfully capture the pollutant pathways.

In order to apply the optimization model to unconfined aquifer problems, we need to understand restrictions and computational approaches. Some formulations lack provisions to limit the optimization search to extraction rates that prevent the physical and numerical dewatering of portions of the aquifer, making it inadequate for solving design problems under many unconfined aquifer conditions. Physical factors like narrow saturation depths or low permeability sediments might cause a well extraction to exceed the aquifer flow into the well. The extraction process eventually ends when the well dries out. The cell is considered to have dewatered, and the simulation model is unable to converge to a useful solution. The dewatering of a grid cell causes an abrupt change in both the head and particle pathways, which leads to a numerical discontinuity in the optimization model. This discontinuity may cause a gradient-based optimization search procedure to fail. Therefore, the optimization technique needs to include the accurate stages to account for any potential numerical issues that may arise while simulating extraction pumping in unconfined aquifers. Constraints on the minimum saturated thickness at numerical grid cells or upper limits on extraction rates can be used to prevent dewatering the aquifer. The latter strategy allows the system behavior to set the upper pumping limits, hence pushing the system's physical limits. However, with this later method, caution must be used to guarantee that the optimization search strategy is still possible in light of the saturated thickness limits.

6.5. CONSTRAINTS

Lower head bound constraints are analogous to minimum saturated thickness constraints in unconfined aquifers, where head is a nonlinear function of pumping. The nonlinearity in an optimization problem with such constraints can be handled by treating it as weak and assuming that head varies linearly with pumping. This approximation is probably sufficient when drawdown is small in relation to the overall saturated thickness. To reflect the current status of the system, linear approximations of the head variations must be updated on a regular basis if drawdown is substantial, as this can lead to severe nonlinearity. Using nonlinear programming techniques is an additional method of accommodating the nonlinear head variation. The goal of the advective control model, which combines simulation and optimization, is to ascertain the lowest possible total pumping required to containing a given pollutant plume. The plume control problem is simulated using two reference frames: (1) forward tracking explicitly represents plume transport, and (2) reverse particle tracking is used to determine the capture zone of a specific groundwater extraction well. Both strategies are used by the simulation optimization model in order to benefit from unique numerical features of each strategy. A collection of stationary control points that are required to be located inside the extraction well capture zone serve as a representation of the plume during the primary stage. In the next step of the solution process, forward tracking is employed, and an extraction well is used to confine all of the particles that constitute the contamination plume to exit the simulation model domain. To ascertain the system state under a set of applied stresses, the simulation-optimization model integrates the USGS simulation models MODFLOW and MODPATH to provide more in-depth analyzes of the simulation-optimization model and model parameterization (Zeinalie et al., 2021).

Establishing a set of sites where a well can be installed is a common optimization-based solution to the pump-and-treat design task. The management model must ascertain the proper pumping rates from this group of well sites, which are referred to as candidate wells. Assuming that well installation costs are low in comparison to operating expenses for long-term plume control devices is another popular strategy for streamlining the optimization problem. Assuming this, a continuous function of the pumping rates can be used to describe the cost function that has to be minimized. It is important to take precautions while simulating unconfined aquifers. Either upper limits on extraction rates or lower limits on head at the well might be placed on the pumping solution if it is considered that the sites of the extraction wells are more susceptible to dewatering. Setting clear upper limits for extraction could unintentionally exclude the best options. In light of this, the strategy can set lower head limits at each potential well location without placing clear limitations on the rates of extraction at those wells.

The advective control model looks for a pumping scheme in which the capture zone completely encloses all control points that represent the contaminated plume during the initial stage of the process. Particles from extraction wells are tracked backward through the velocity field to replicate the capture zone. A distance measure is used, where the minimum distance between each plume control point and all particles is constrained, to numerically reflect the plume capture limitations. The plume control point is located inside the capture zone when the distance between it and the particle pathway is equal to zero. All control points must have a constraint function of zero in order to guarantee capture of the whole plume. Many candidate wells and several particles in confined and unconfined aquifers can be encountered in two- and three-dimensional plume capture problems that can be resolved by the particle tracking-based advective control models. The design makes use of both forward and reverse particle tracking to take advantage of the properties of the two reference frames and offers a simple method for handling plume management concerns. Aquifer dewatering can significantly reduce the feasible region of the optimization problem in some circumstances, such as an unconfined aquifer with high extraction rates or a small saturated thickness. By adjusting the search method, the impacts of aquifer dewatering and the nonlinear head response to pumping are taken into account. The computing cost increased significantly as a result of the extra algorithmic steps.

6.6. POLLUTANTS

One strategy for developing pump-and-treat systems for unconfined aquifers could be to find the best solution in confined settings first, then test this solution in unconfined settings to look for cell dewatering. The constrained solution could serve as the starting point for an optimization search utilizing the unconfined assumption if dewatering is not observed. However, if drawdown is large, this strategy is probably going to produce a solution that is not workable in unconfined environments. Pollutants are transported toward extraction wells by natural flow patterns made possible by the pumping solutions. After that, the main purpose of the active wells is to modify gradients in the plume's downgradient. Many of the particles remain in the aquifer for as long as feasible while being managed by the pumping pattern, which is a result of the advective control design. Higher pumping rates than what is considered appropriate will typically lead to higher velocities. This enables the pollutants to be affected by natural processes like biological degradation before they are brought to the surface for treatment (Saravanan et al., 2021). This method might work well for dispersed, low-concentration parts of a plume, but it might not be the best course of action for hotspots, or places of high concentration. Alternatively, wells positioned inside or close to the high concentration area may be required to regulate hotspots. It is advisable to look into methods for limiting the hot-spot areas that a subset of the candidate wells can collect. The extensive pollution of water bodies by halogenated and unsaturated hydrocarbons (pesticides and solvents) has prompted investigation into several remediation solutions. With several benefits over alternative methods, treatment by reductive dehalogenation and hydrogenation with Pd catalysts exhibits potential. Pd catalysts can reduce a wide range of substances in aqueous solution at room temperature, including format/carbonate, oxidized nitrogen species, PAHs, and hydrogenated alkanes, ethylene, and aromatics. Halogenated intermediates, the halogenated compounds often react to generate the corresponding simple alkanes.

For a very long time, catalysts have been employed in a wide range of organic processes, such as hydrogenolysis, oxidation, reduction, exchange, hydrogenation, and dehydrogenation. Some works focused on aqueous phase hydrogenation and reductive dehalogenation processes mediated by palladium (Pd) catalysts. Environmental engineers who are trying to clean waters contaminated by halogenated and/or unsaturated hydrocarbons may find these interactions interesting. Typically, catalyst particles are made of a metal that has been deposited on a support surface. Pd is frequently reported to be the most active metal among those employed for catalysis. Pd catalysts considerably outperformed Pt or Rh catalysts in the aqueous hydro-dechlorination of 1,1,2-trichloroethane. The shape, size, porosity, and surface area of the catalyst can vary. Common materials include carbon, alumina, silica, and zeolites.

The majority of catalytic research in the past has been done on organic or gas phases. On the other hand, research on the use of supported palladium catalysts and a hydrogen source to treat organic molecules in the aqueous phase at room temperature has surfaced recently. Halogenated hydrocarbons frequently pose a risk to human health and are resistant to treatment. Granular activated carbon is used in fixed bed reactors as part of the conventional treatment process to absorb pollutants from river streams' however; this leaves behind contaminated solids that need additional treatment.

Many partially and fully hydrogenated compounds are produced via a convoluted series of hydrogenation processes that involves phenanthrene. The detected products did, nevertheless, seem to attain equilibrium concentrations. One of naphthalene's two aromatic rings was hydrogenated to create 1,2,3,4-tetrahydronaphthalene. The kinetics of catalytic transformations in reactor systems are a complicated function of many factors, including catalyst parameters and reactor configuration and operation. For instance, the amount of Pd/Fe supplied to the system, Pd loading on the Fe, and the composition of the aqueous solution all affected the rate of PCB hydrodechlorination on Pd/Fe. Comparing data from different research is quite challenging due to the complex nature of the kinetics' nevertheless, comparisons within a single system might provide valuable information. Some researchers employed several reactor setups and catalysts, and reported first order reaction rate constants for chlorinated ethylene's' the values varied, but the patterns within each set of tests were comparable. As the process is scaled up from laboratory systems to field units, it is also reasonable to predict that similar tendencies will stay true, albeit with significant system dependence in the absolute rates of response. Direct electric currents can be designed to mix pollutants, bio-stimulants, and even bioaugmentation inoculants to improve in situ bioremediation. These changes in biochemistry and electrochemistry have an effect on the subsurface environment. For anaerobic organisms, electric field intensities above 1.5 V/cm appear to cause an environmental shock. When the electric currents are turned off, these cultures can resume their activity. It appeared that aerobic cultures were not negatively impacted by field intensities lower than 0.28 V/cm. During the first 24 hours of exposure, higher DC field strengths, up to 1.14 V/cm, seem to encourage aerobic cultures. Nevertheless, after the first 24 hours, prolonged exposure to dc fields (1.14 V/cm) appears to slow the growth of aerobic cultures (Alshawabkeh & Maillacheruvu, 2002).

6.7. WASTE

Many hazardous waste sites have been found to include organic pollutants. One major issue is the cost-effective restoration of these contaminated areas to levels that are acceptable to the environment. The efficacy of current in situ soil remediation technologies is dependent on air and hydraulic flow, and they are less effective in cleaning up lower hydraulic conductivity soils, such as fine sands, silts, and clays. A frequently affordable solution for treating soil and groundwater contaminated with organic materials is in situ bioremediation. The effective infusion of nutrients and electron acceptors into the porous medium is a prerequisite for the successful application of in situ bioremediation. An electron acceptor, macronutrients (phosphates and nitrogen), micronutrients, trace nutrients, and an electron donor are needed for microbial activities. Co-metabolite reactions mediated by microbes also need an additional external electron donor. In rare situations, the microbial transformation reaction might require inducers to start. Low soil permeability, preferential flow pathways (channeling), biological uptake, and chemical reactions in the soil prevent these additives from being introduced and transported effectively. A nutrient-rich area with excessive biological development may arise from flaws in the hydraulically-driven transport systems (biofouling). System implementation is negatively impacted by biofouling because microbial growth clogs the flow pathways, reducing conductivity. The main reason why certain in situ bioremediation attempts fail is improper remediation transportation. The use of electric fields to move bacteria and nutrients across hazardous waste sites is an emerging technique for treating them utilizing in situ bioremediation techniques. In order to get around issues with injecting chemicals into soils that are heterogeneous or have low permeability, electric fields could be utilized. Electric injection, as opposed to hydraulic injection, can provide nutrients and electron acceptors in a uniform manner. The pore fluid charge and ionic mobility govern the uniform movement of ions under electrical fields. Electric fields can be used to accelerate and uniformly transfer nutrients, electron acceptors, and microbes with lower hydraulic conductivity or heterogeneous soils. To improve additive injection, the method could also introduce electroosmotic flow into fine-grained soils. Both aerobic and anaerobic bioremediation can be accelerated with this technique.

The geochemical (soil) and electrochemical processes may be impacted by the applied electric field (Alshawabkeh et al., 2004). To know the potential of injecting and transporting nutrients and bio-stimulants by ion injection and electroosmosis, a review of the ion transport mechanism in porous media under dc fields is needed. The effects of electrolytic processes on pH and dissolved oxygen (DO) levels, which in turn influence microbial life must be considered. Electrolysis reactions are significant because they generate hydrogen and oxygen and have the potential to degrade pollutants abiotically. The variables that could affect microbial adhesion and transit in dc fields are important. Electrokinetic transport processes in an ion exchange medium regulate the injection and transport of bioremediation additives by electric fields. Electromigration, also known as ionic migration, electroosmotic advection, electrophoresis, and, to a lesser extent, diffusion is some of the transport modes. While electromigration is the separation of anions (negative ions) and cations (positive ions) by their migration to the anode and cathode, respectively, electroosmosis is the mobilization of the pore fluid in the soil, typically from the anode (positive electrode) toward the cathode (negative electrode). Since the ions can migrate in a stationary pore fluid, electromigration does not require fluid flow. The process of transporting charged particles, like bacteria or clay particles, to an electrode with the opposite polarity under a dc electric field is called electrophoresis. When the components of diffusion, ionic migration, and electroosmosis are added, the total mass flux of a charged species or particle, which can act as a donor or acceptor of electrons, is calculated.

6.8. PH

Complex boundary conditions that are created at the electrodes by oxidation and reduction can either speed up or slow down the breakdown of contaminants and microbial activity. Aerobic deterioration is stimulated by oxygen produced by anode oxidation. Also, an acid is created, which has the potential to lower the anode pH to less than 2. The acid may acidify the soil as it travels toward the cathode under the applied electric field. The soil ability to buffer pH variations determines the soil tolerance to such changes. A pH reduction, however, is not ideal for the growth of microorganisms. The ideal pH for bacterial growth is close to neutral, with minimum and maximum pH values for growth being close to 5 and 9, respectively. For biological activity in the soil to be maximized, the pH level must be generally neutral. Mineral breakdown in the soil can also result from soil acidity. It is important to take into account electrolyte conditioning that reduces the production of acid and improves the supply of the correct chemicals for bioremediation. For instance, it was successful to neutralize the acid and base at the anode and cathode and inject ammonium and sulfate ions into the soil by using ammonium hydroxide at the anode and sulfuric acid at the cathode. Another crucial boundary condition that is produced at the anode is the generation of oxygen. To improve aerobic conditions, the anodic oxygenated water can be efficiently injected into the soil through electroosmotic flow.

The efficiency of electrolysis in producing gas is 75%. To increase the amount of oxygen required, additional methods can be employed, such as air sparging, in-well aeration, hydrogen peroxide, oxygen-releasing solid compounds, or even cryogenic oxygen generators. On the other hand, DC fields have the benefit of being able to generate and transport oxygen in soils with low hydraulic conductivity. It is important to remember that although this oxygen will improve the environment for aerobic microbes, the anaerobic species may die as a result. Anaerobic microorganisms may be negatively impacted;

therefore if anaerobic bioremediation is the intended treatment approach, steps must be taken to remove oxygen at the anode using chemical scavengers or gas stripping. As an alternative, an aerobic-anaerobic treatment might be developed to benefit from the oxygen present in a small subsurface zone. Contaminants have been removed from underground using electric fields. Electric fields have been shown to be effective in extracting radionuclides and heavy metals from soils through lab scale tests and limited field studies. The electroosmotic extraction of organic pollutants from soils has also been accomplished using electric fields (Cameselle & Reddy, 2012).

6.9. ELECTRIC FIELD

The application of electric fields in soil restoration has primarily concentrated on the extraction of contaminants by ionic migration and electroosmosis. One effective method for eliminating ionic or mobile pollutants from underground areas is by the use of electric fields for contamination extraction. Nevertheless, this method may not work well for treating soils that have trapped or immobile organic materials, such as dense non-aqueous phase liquids (DNAPLs).

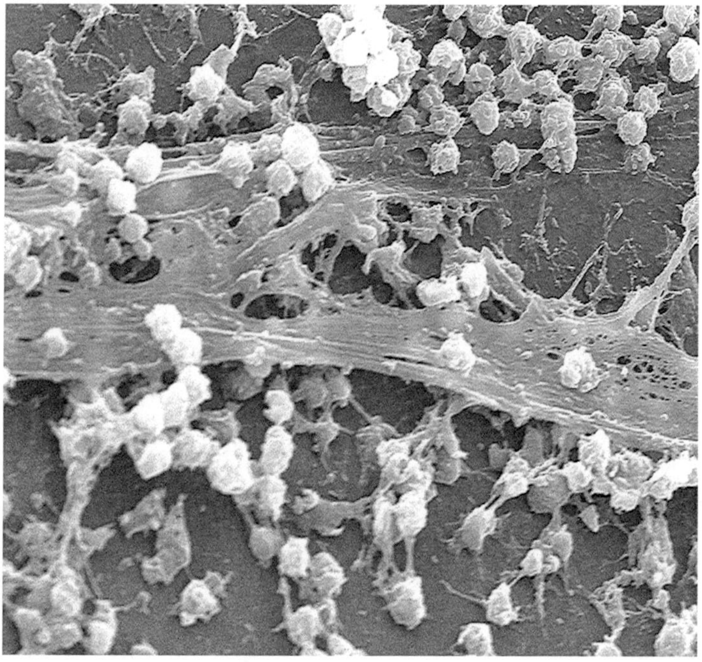

Figure 6.1. *Biofilms.*

Source: https://en.wikipedia.org/wiki/Biofilm#/media/File:Staphylococcus_aureus_biofilm_01.jpg, Creative Commons License.

Electric fields can be used to promote in situ biodegradation of these organic materials in either anaerobic or aerobic situations. Before using the method, it is required to assess how dc electric fields affect the biogeochemical interactions. The effect of dc electric fields on microbial adhesion and movement in the subsurface is still unclear. Furthermore, nothing is known about how dc fields affect the activity of microorganisms within a soil matrix. It has long been known that the presence of suitable microorganisms at the actual site of contamination is crucial for determining whether biodegradation or biotransformation will occur and for affecting the rate of biodegradation. Microorganisms can exist as biofilms, microcolonies, or in suspension in the pore fluid below the surface.

6.10. MICROORGANISMS

The extent to which a subsurface medium is colonized is frequently influenced by bacterial adherence to porous media. Surface conditioning, microorganism transportation to the conditioned surface, and microorganism sorption to the surface are the three steps in the formation of biofilms (Figure 6.1). The research on biofilms acknowledges that the characteristics of the bacterial cell surface have a crucial role in defining the bacterial ability to adhere to various media and, ultimately, the architecture of the biofilm. Microorganisms sorb via reversible sorption, which is controlled by charge/electrostatic interactions, and irreversible sorption, which is caused by the synthesis of extracellular exoploymers and matrix formation. An applied electric field may be important for bacterial adherence and movement in subsurface environments, as evidenced by the reversible sorption of bacteria on charge/electrostatic interactions. Furthermore, the surface characteristics of bacteria, such as their lipid content and surface charge, may have an impact on how easily they can move electrophoretically across porous substrates. Data suggest that the presence of heavy metals affects the electrokinetic characteristics of bacteria. It is anticipated that the applied dc electric field will also have an impact on the adhesion, transport, and electrokinetic characteristics of microorganisms. The microorganisms have a tendency to adhere to the electrodes. It was not assessed what kind of microbes were affixed to the electrodes or how active they were. Since microorganisms were bonded to both the anode and the cathode, electrode polarity did not appear to have an impact on adhesion. This could suggest that the positively charged anode and the negatively charged microorganisms are not attracted to each other electrically, which could explain the adhesion to the electrode surface. To confirm whether this adhesion is caused by electrode charge, irreversible sorption, or reversible sorption, more analysis is required.

In addition to influencing microbial adhesion, electric fields also have an effect on how well they move across porous substrates (Or et al., 2007). Microbes will be transported towards the anode by dc fields because they are often negatively charged. In general, soil physical characteristics like porosity and pore size distribution have influence on the rate of transport, which is correlated with the effective electrophoretic mobility in soils. It is also important to remember that, as living things, microbes may be susceptible to other factors in addition to their propensity to form colonies and adhere to the surface of soil particles. As such, the movement of microorganisms in porous media under an electric field may not be solely controlled by electrophoresis. At different electric field strengths, the effects of electric current on the surroundings and the anaerobic microbial activity in batch reactors were investigated. Bio-electrokinetic (BioEK) reactors with mesh electrodes installed at both ends that are coated in titanium, are used for conducting experiments. There were applied electric fields ranging from 1.5 V/cm to 6 V/cm. These studies employed unacclimated anaerobic cultures that were taken from a mesophilic anaerobic digester. If the water electrolysis reactions proceed at the same rate and efficiency, there might be limited pH variations. The hydroxyl ion produced at the cathode in a mixed reactor should be neutralized by the proton generated at the anode. Even in batch reactors with fully mixed conditions, the data showed that the pH dropped to less than 5.5. The pH drop indicates reduced hydroxyl synthesis at the cathode, which could be attributed to biological events occurring in the reactor or to various electrolysis reactions. The type and quantity of ions in the solution will affect the pH variations, necessitating more research. The system was effectively buffered for the range of electric field intensities evaluated by the use of sodium bicarbonate.

When no oxygen scavengers were used, the experimental reactors produced dissolved oxygen in proportion to the applied electric field intensity. However, the system's dissolved oxygen was managed by adding sodium sulfite. Even though some dissolved oxygen was created and not entirely removed using oxygen scavengers, anaerobic bacteria may be able to thrive in oxygen-free niches found in normal soil. In fact, some anaerobic bacteria may have survived in the presence of high oxygen concentrations, especially sulfate-reducing bacteria, or SRB. Over an exposure period of approximately 140 hours, the electric current in the reactor generally raised slightly with time at different electric field strengths. The results of this preliminary investigation show that microbial activity and the efficiency of removing dissolved organic carbon (DOC) do not seem to be significant, especially at the lower end of electric field strengths, even though this suggests that the ionic composition of the electrolyte medium changed over time. The ability of the culture to grow following exposure to electric currents was used to evaluate microbial activity. The ability of the anaerobic bacteria to consume readily degradable acetate (DOC) was used to calculate the microbial activity (Vorkamp et al., 2001). At regular intervals, a culture sample was taken out of an experimental reactor that was fed a batch. A portion of the sample was subjected to rapid analysis and designated for DOC concentration, while the remaining portion was left to recover in a glass vial for a whole day in order to maintain anaerobic conditions. The DOC was assessed once more after a 24-hour period.

The microbial activity data for experiments conducted at 1.5 V/cm and 4.5 V/cm electric field strengths were recorded. These results suggest that prior to the application of the electric current; the sample had a higher level of DOC removal (20%) than the sample. It is clear that the sample exhibits less activity than the sample after the initial shock exposure to the electric current. For several hours, this pattern persisted, and throughout that time, all of the examined total DOC percent removal in the experimental reactor decreased by roughly 10% to 20%. Even with the application of the electric field, the sample gradually displayed more DOC removal than the control after a few hours of exposure, indicating an increase in microbial activity. This is an interesting phenomenon because it suggests that the culture has become accustomed to electric current and that the anaerobic culture has a propensity to recover from the initial shock load in terms of environmental variables that have changed as a result of the application of the electric current. The experimental reactor rate of DOC percent elimination improved steadily after the electric current application was terminated. When exposed to electricity, full aeration and mixing kept the dissolved oxygen close to saturation. During the tests, there was no discernible change in the temperature. Sludge mixing minimized large pH changes by allowing electrolytic reactions at the electrodes to neutralize one another, despite the minor pH fall (Zeng et al., 2022). Anaerobic testing revealed a pH decrease to roughly 5.5. The difference in processing time and the components of the anaerobic and aerobic cultures may be the cause of the small pH reduction in the aerobic tests as compared to the anaerobic tests. It is probable that an additional electrolysis reaction, apart from water reduction, took place at the cathode during the anaerobic testing, thereby restricting the formation of OH^-. Since temperature, pH, DO, and other variables were all under control, the only factor that might be causing variations in microbial activity was electric currents. The sludge percent change in chemical oxygen demand (COD) was utilized to measure the amount of microbiological activity. Additional substrate utilization metrics, like TOC, DOC, BOD, and/or VSS, may be applied. Nonetheless, COD was employed because of its feasibility and ease of use compared to certain other metrics. Because bacteria are organic in nature and COD may not correctly indicate microbial activity, some difficulties may arise (Rozzi & Remigi, 2004). COD levels are a reliable indicator of microbial activity.

To keep the pH, DO, and temperature constant, the tests were carried out in an open, mixed, and aerated reactor. Therefore, pH and temperature may not have any bearing on the variation in COD drop. In all tests, aeration and mixing kept DO close to saturation, minimizing the impact of oxygen generation at the anode. The abiotic transformation caused by electrolytic reactions at the electrodes is the only other process that could be connected to COD decline. Increasing the current density should cause the COD drop to increase if abiotic redox of the organic material occurs. The primary cause of the variation in COD decrease observed in tests with varying current densities might be attributed to microbial activity. Because of the poor mobilization of the contaminants and the difficulties in delivering chemicals uniformly, contamination in low permeability soils presents a considerable technical obstacle to in situ remediation. Trichloroethylene (TCE) is mobilized in soil through an alternate method called electroosmosis (EO). Nevertheless, the EO method significantly alters the chemical composition of the soil, which could have an impact on TCE transport and/or chemical transformation. There was a contamination zone (CZ) one centimeter below the anode. From anode to cathode, there was a vertical downward flow of electroosmotic fluid. For four weeks, a voltage gradient of 1.4 V/cm was applied to the soil. Over the course of 672 hours, more than 95% of the TCE was mobilized in the soil toward the cathode. Dichloroethylene (cis-1,2-DCE), which shows that TCE has been dechlorinated, was observed (Lin et al., 2018). In areas of the soil column where reducing conditions predominate, dechlorination took place. TCE underwent reductive dechlorination in the vicinity of the cathode, which serves as an electron source during electroosmosis. The findings indicate that the remediation efficiency during EO may be improved by the possible chemical transformation of chlorinated organic molecules.

6.11. ORGANIC POLLUTANTS

In the US and other countries, organic pollutants are frequently found contaminating soil and groundwater. Groundwater rapidly transports chlorinated hydrocarbons, which represent a chronic environmental threat. Trichloroethylene (TCE) is an organic solvent that has been chlorinated and is extensively utilized in various industries such as paint, electronics, dry cleaning, and metal processing. TCE has entered many subterranean habitats as a result of spills, leaks, and dumping. Because TCE and related chlorinated alkenes are thought to be carcinogenic and typically do not biodegrade in the environment, they pose a severe risk of groundwater contamination. Pump and treat have been the most common method used to try to clean up aquifers that contain TCE. This involves either pumping the contaminated water or vacuuming the contaminated air to certain above-ground treatment facilities. Pump and treat method is restricted by hydraulic conductivity, preferred flow, sorption onto soil particles, and residual saturation. Many in situ soil remediation processes require moving contaminants to a treatment/collection zone or delivering nutrients, microorganisms, and chemicals to degrade the contaminant. In general, Pump and treat techniques have limited removal capability, and after pumping is stopped, the contaminant concentrations might increase again. It is possible to transport treatment solutions hydraulically to the contaminated zones in soils with high hydraulic permeability. Degradation products and mobilized pollutants can be eliminated in the same way. Pump and treat techniques have proven effective in sandy soils, such as aquifers that provide large amounts of water. On the other hand, hydraulic conductivity, soil matrix effects, desorption, and residual saturation of TCE limit the removal effectiveness, making pressure-driven hydraulic delivery and removal of pollutants in low permeability soils unfeasible.

6.12. ELECTROOSMOSIS

Electroosmosis (EO), a process that involves applying an electric current to the soil, is one way to move chemicals and solutions in poor permeability soils. Ion mobility in the double layer of clay surfaces produces EO fluid flow. Because of this, EO works best in soils that are rich in clay and have tiny particles. Similar to how the rate of hydraulic flow is proportionate to the hydraulic gradient, the magnitude of electroosmotic fluid flow is proportional to the applied electrical gradient. Therefore, organic pollutants may be carried by EO, whereas electromigration may be used to move dissolved ions and other charged species. Recently, this approach has shown promise for effective, economical in-situ remediation of both organic and inorganic pollutants in fine-grained soils. It has been widely used in engineering applications. Heavy metals like copper, zinc, cadmium, and lead have all been effectively mobilized using EO. For substances including benzene, toluene, xylene, phenol, and chlorinated solvents, organic pollutant removal has been observed.

CHAPTER 7

SUSTAINABLE GROUNDWATER MANAGEMENT

Contents

7.1. INTRODUCTION

A common inorganic contaminant in shallow aquifers is nitrate. Industrial effluents, municipal wastewater systems, and agricultural fertilizers can all lead to nitrate contamination. Nitrate concentrations in aquifers vary both geographically and temporally. This variability is typically caused by changes in the direction of groundwater flow and nitrate reduction.

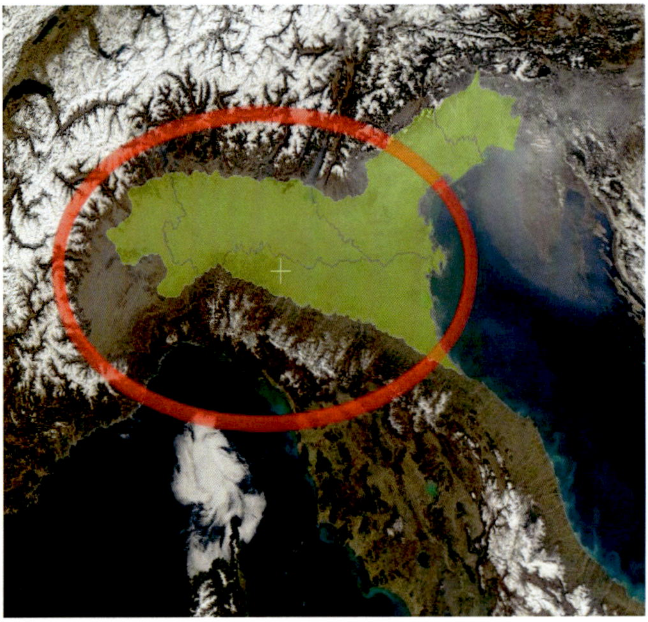

Figure 7.1. *Po River valley.*

Source: https://en.wikipedia.org/wiki/Po_Valley#/media/File:Pia_padana.jpg, Creative Commons License.

The Po River valley (Figure 7.1) in Italy is the largest and most heavily farmed alluvial plain (Marchetti, 2002). It is severely affected by surface water eutrophication and nitrate contamination. However, depending on the nitrogen load and subsequent denitrification, different crops, irrigation techniques, soil tillage, and nitrogen fertilizers form infrastructure for agricultural operations. Heterotrophic denitrification, which produces nitrogen gases by using NO^{3-} as an electron acceptor and a carbon source as an electron donor, is a quantifiable process of nitrogen reduction from surface and groundwater systems. Several studies examined the flow of nitrate from the upper soil to the groundwater in farmed areas and assessed if adding chicken manure from organic farming is a feasible substitute to reduce the NO^{3-} content in shallow groundwaters. The identical treatment rates were used in separate locations for testing purposes, but the soil textures were different. By adding additional organic matter, chicken manure was used to improve the fertility of the sandy soil. The release of organic substrates from the mineralization of manure was adequate to stop nitrate leaching in the sandy soil. Consequently, some studies highlight how crucial it is to take into account how organic matter functions as a buffer for NO^{3-} excess through denitrification reactions. Adding chicken dung to the soil organic matter gave rise to a supply of organic acids, such as acetate, which are utilized as substrates for the denitrification process. Conversely, in the unaltered silty loamy soil, the persistent lack of organic matter was unable to stop nitrogen from seeping into the shallow unconfined aquifer. A relatively low but continuous release of organic substrates was found to be adequate to prevent nitrate leaching, even in the vulnerable area (sandy soil) to prevent groundwater pollution. However, more investigation should focus on the type of manure applied and the farming methods i.e., industrial or organic. Because of the presence of hormones, antibiotics, and other undesirable substances that may affect the quality of groundwater and interfere with bacterial activity, this may affect the kinetics of organic matter transformation and the rates at which nitrogen mineralizes. The use of $_{15}N$ and $_{18}O$ nitrate isotopes to identify the source of nitrate and mathematical modeling to measure the mass transport of nitrate leaching is becoming popular (Matiatos et al., 2023). The isotopic and mathematical modeling techniques are essential instruments for monitoring the subsurface fate of nitrogen species.

7.2. MODELING

Integrated modeling that accounts for soil, unsaturated, and saturated zones is crucial for developing effective management strategies to reduce groundwater nitrate pollution. Regional studies on groundwater management frequently overlook the impact of the unsaturated zone. The EU Water Framework Directive requires good groundwater quality, and the unsaturated zone can have a significant impact on the time taken for nitrate transport. Researchers combined a basic unsaturated zone model with a hydro-economic framework to determine the optimal rate of fertilizer application, both spatially and temporally, to maximize net agricultural benefits while adhering to groundwater quality criteria at different control locations. The aquifer El Salobral-Los Llanos (Spain) was subjected to the integrated model. Some findings indicate that, when taking into account the unsaturated zone, a delay of roughly seven years can be anticipated in meeting the quality standards' nonetheless, the overall benefits remain relatively unchanged. Aiming to establish an integrated management framework for sustainable water use, the EU Water Framework Directive mandates that all water bodies attain a good (in terms of quality) status by 2015. Consequently, it is crucial to accurately estimate how long it will take for groundwater bodies to reach a high standard of quality. In order to mitigate diffuse contamination into groundwater, management decisions must be analyzed and put into action. We can determine the response time in order to forecast future values of groundwater quality, particularly following the implementation of environmental measures like cutting back on fertilizer use. The soil parameters, the degree of saturation water, which is regulated by the variable water input at the land surface, and the distance between the source area and the control site, where water quality is analyzed, all affect the response time between the application of fertilizer and contamination at a specific site. Advective-dispersive transport will continue to move nitrate across the unsaturated and saturated zones while denitrification mostly takes place in the root zone (Mekala et al., 2017). Hydro-economic model was used to determine the best distribution of fertilizer application to maximize agricultural benefits while keeping groundwater nitrate concentrations below a certain standard.

Agronomic, flow, and transport models were combined in the hydro-economic model inside an optimization framework. Quadratic functions that link the amount of water and fertilizer used, crop production, and nitrate leached are obtained from agronomic simulations. The output of agronomic simulation models, such EPIC, can be used to generate results. The concentration response matrices, which display the impact of a pollution source on nitrate concentrations at several control sites over time, are derived from the groundwater flow and nitrate transport models. The groundwater nitrate concentration development at several control sites over time, resulting from many pollutant sources scattered over time and space, can be simulated by superimposing the response matrix integration with the management model restrictions. Crop prices and agricultural production functions were used to calculate the benefits of agriculture. The hydro-economic model was combined with a basic unsaturated model that solely takes gravity flow into account. By allocating solute concentrations to the inflow rates, a basic solute transport model can be generated. This approach ignores the concentration of immobile solutes by only accounting for the mobile portion of the porous media. By integrating the solute mass as it exits the soil column throughout time and the solute mass at the appropriate outflow timings, as well as by computing the solute mass and volume increments to produce outflow concentrations, irregular outflow concentrations can be obtained. The majority of nitrogen cycle events occur in the root zone, where nitrate transport is thought to be conservative in the saturated zone. Although the fertilizer needs to be further lowered when accounting for the nitrate transport time across the unsaturated zone; this reduction did not significantly lower the overall advantages.

7.3. POLLUTANT REMOVAL

The amount of recharge and the wet-dry phases have the most effects on the transport time across the unsaturated zone. Different time delays can be anticipated if there are changes in land use patterns and/or precipitation patterns. This emphasizes how management strategies to lower groundwater nitrate pollution must take into account the unsaturated zone. Based on field research at a dairy farm in Saxony, Germany, a field test strategy for subsurface iron removal was implemented (Kivits et al., 2018). A tracer test was carried out at the field location to measure the dissolved oxygen concentration and electrical conductivity in order to ascertain the size of the reaction/response zone. The test results are utilized to optimize the performance of the aeration units and wells, particularly when contemplating long-term subsurface treatment with oxygen or aeration. Anoxic groundwater contains iron and manganese. Although high iron concentrations do not pose a health risk to people or animals, they can cause several problems, including production well blockages (clogging), incrustation in water supply distribution systems, precipitation, orange or brown water, and unpalatable drinking water. Two approaches can be used to remove iron from groundwater. The primary factors in selecting one of the groundwater treatment methods are the raw water quality, the available funding, and the methodology. After abstraction, the most popular procedure is treating groundwater on the surface. Subsurface iron removal can be a practical and affordable groundwater treatment method under some boundary conditions. Iron hydroxides are produced in the aquifer surrounding an abstraction well by oxygen-enriched water seeping through it, creating adsorption sites for Fe (II) and other cations. Water devoid of iron is ultimately produced through a periodic cycle of abstraction and infiltration.

Another chemical-free, environmentally friendly in-situ oxidation method is subsurface iron removal. The subsurface soil matrix serves as a naturally occurring bio-chemical reactor for this technique. Subsurface iron removal is a process that is effectively used by over 10,000 people in Europe. Drinking water production ranges from 1 m^3/d to over 30 000 m^3/d (Bodzek et al., 2011). Additionally, there is a large range in the dissolved iron and manganese contents in the groundwater, 0.5 to 55 mg/l for iron and 0.1 to 3.0 mg/l for manganese. Subsurface iron removal has been demonstrated to be a practical, affordable, and sustainable method of treating drinking water. However, its potential is not being fully realized since waterworks are not aware of this fact. Both small- and large-scale applications of subsurface iron removal are possible, as well as combinations with other treatment methods. The amounts of nitrite and sulfide determine how much oxygen they absorb. High concentrations of Fe (II), Mn (II), and ammonium, as well as low pH and hardness, are conditions that make subsurface iron removal difficult. A pH level higher than 5.5 is necessary for the oxidation of Fe (II) to Fe (III). Depending on the water-soil system buffering ability, the pH drops during iron oxidation. It is helpful to conduct a subsurface iron removal field test and groundwater sample prior to designing and constructing an in-situ treatment plant for iron removal in order to determine the functioning principle with the chemical composition of the natural groundwater. At the location selected for the in-situ treatment, at least one monitoring well is required for both investigations. The volumetric ratio of the water abstracted to the water infiltrated, which is typically between 3 and 5 but can exceed 10, determines the efficacy of the subsurface iron removal procedure.

7.4. ABSORPTION

It is also feasible to eliminate dissolved organic carbon, arsenic, and manganese through adsorption in the reaction zone. Iron adsorption on soil grains occurs during groundwater abstraction in the first month, but iron oxidation in this area surrounding the well must be completed before manganese adsorption and oxidation may occur. A pH-Eh curve indicates that iron and ammonium must first undergo oxidation before manganese can begin to immobilize. Manganese will eventually adsorb onto the soil grains if there is enough space for iron and manganese precipitation around the well. The autocatalytic activities of microorganisms also contribute to the acceleration of manganese oxidation. DNA-microarray was used in detecting and characterizing the microbial populations engaged in the degradation of organic contaminants in contaminated groundwater plumes. The 5' terminus of the 16S rRNA genes for Bacteria and Archaea was amplified from groundwater DNA using universal PCR primers (Gantner et al., 2011). In order to identify the Bacteria and Archaea families simultaneously, the 78 16S rRNA gene targeted oligonucleotide probes in the prototype microarray were hybridized with the amplification products after being fluorescently labeled. The findings show that DNA microarray technology makes it easier to quickly and effectively analyze diverse microbial communities in environments that are difficult to investigate, including contaminated groundwater. Hence, natural attenuation (NA) in contaminated groundwater can be monitored using DNA-microarray technique. The primary process of natural attenuation is the breakdown of organic pollutants by microbial populations in soil and groundwater. By using DGGE of PCR-amplified marker gene fragments like the 16S rRNA gene, the impacts of pollutants on the growth characteristics and composition of microbial communities have been confirmed. For the purpose of comparing the microbial communities engaged in the degradation of the wide range of pollutants, PCR-DGGE has been the preferred approach. PCR-DGGE is a common approach in microbial ecology, however it has some important drawbacks. It is completely site-specific and makes it difficult to precisely monitored natural attenuation since it makes it difficult to identify the microbial communities responsible for the breakdown of pollutants. The science of microbial ecology has advanced significantly in recent years. By specifically detecting microbial rRNA genes, the use of DNA microarrays in the fields of marine biology and soil- and water-ecology allowed the analysis of microbial communities at the species level. It was assumed that NA in contaminated groundwater may be monitored using DNA microarray technology. One of DNA-microarray advantages over other state-of-the-art molecular biology approaches is its capacity to analyze many parameters simultaneously.

7.5. MOLECULAR TECHNIQUES

Any taxonomic level can be used to assess the composition of microbial communities using known probe-sequences. In order to address the drawbacks of PCR-DGGE and use the advantages of DNA-microarray technology in environmental analytics, a 16S rRNA directed DNA-microarray for the examination of microorganisms found in soil and groundwater was produced (Malik et al., 2008). The primary goal of microarray design was to use the fewest number of probes to detect the highest number of bacteria. Regular monitoring of natural attenuation cannot be done using a microarray for the detection of microbes on a species level because over 90% of the known prokaryotes are still unknown. The DNA-microarray developed made it possible to simultaneously detect Bacteria and Archaea families in groundwater. A sharp decline in pollutant concentrations was observed down gradient from the landfills under investigation as a result of natural attenuation. Organic molecules with a relatively high molecular mass are the primary contaminants that seep from landfills. Increased concentrations of NH_3, H_2CO_3, and CH_4, which result from the breakdown of organic molecules, serve as a sign for the bacterial breakdown of pollutants. Anaerobic groundwater makes SO_2^{4-}, a major supply of electron acceptors for anaerobic respiration, which is a necessary condition for anoxic carbon metabolism. The level of contamination is directly correlated with the concentration of bacteria in groundwater samples. This suggests that the contaminants emerging from the landfills do not prevent microbiological development. Microorganism concentrations in polluted samples are roughly five times greater than in neutral samples, and they drastically decline along the groundwater contamination plume. Consequently, high levels of pollutant breakdown are indicated by increasing bacterial concentrations. As anticipated, bands with an average size of 540 bp and 800 bp, respectively, are produced when sample DNA is PCR-amplified using primers unique to bacteria and archaea.

The yields of DNA amplification from groundwater samples analyzed by PCR were found to be proportionate to the amounts of microorganisms measured in the samples using epifluorescent cell counting. The yields of PCR amplification were significantly lower in highly contaminated samples. This discrepancy between bacterial concentration and PCR results is likely the result of inadequate DNA extraction procedure. Organic matter is fragmented down by microbes to produce humic compounds. They can be classified into humin, fulvic acid, and humic acid. They are resistant to additional biodegradation. Humic compounds can chelate positively charged multivalent ions such as Mg^{2+}, Ca^{2+}, and Fe (II), as well as other ions, because of their phenolic and carboxylic acid functional groups (Pan et al., 2011). Humic compounds can chelate Mg^{2+} ions, which can stop the reaction since Mg^{2+} concentration is an essential factor for PCR. The findings suggest that the DNA-extraction procedure used does not make it easier to completely remove the common humic compounds from groundwater samples. Glass slides with epoxy were covalently attached to amino-modified DNA probes for the detection of Bacteria and Archaea families, which have been found in soil and groundwater. The DNA-microarray allowed the simultaneous detection of Bacteria and Archaea. The normalization factor for the hybridization signals specific to Bacteria and Archaea was obtained from the hybridization signals of probes. The relative abundance of Bacteria and Archaea-families in the samples were represented by normalized data, which was obtained by dividing each Archaea- and Bacteria-specific signal by the arithmetic mean of its associated normalization factor. The microarray data was corrected for the concentration of microorganisms found by epifluorescent direct counting, which also made it easier to perform a semi-quantitative evaluation of the microbial communities in the groundwater samples.

7.6. PROBES

Some control probes may not provide any hybridization signal above background/normal level, indicating that the chosen hybridization conditions were stringent. Aerobic microorganisms including the Bacillaceae, Pseudomonadaceae, Neisseriaceae, and Streptomycetaceae are known as mycobacteria. The hybridization signal produced by probes intended to identify various Bacteria-families was below the background level. These findings demonstrate the anoxic conditions in the groundwater samples that were obtained, as well as the taxonomic specificity of the chosen probes. In every groundwater sample that was analyzed, the DNA of sulfate-reducing bacteria from the taxonomic groups Desulfobacteraceae, Desulfobulbaceae, Desulfovibrionaceae, and other sulfate-reducing bacteria was found. A signal for contamination can be found in contaminated groundwater samples because of the increase in the sulfate-reducing bacteria. DNA of methane-producing Archeae from the families Methanobacteriaceae, Methanomicrobiaceae, and Methanosarcinaceae can only be found in contaminated groundwater, in contrast to the widely distributed sulfate-reducing bacteria. A good marker for bacterial degradation of organic pollutants is the drop in TOC concentration in the groundwater samples, which correlates to a decrease in the concentration of sulfate-reducing bacteria downstream from the contamination source. Compared to the rate of TOC, the methane-producing Archaea are decreasing at a substantially faster pace down the downstream flow from the site of pollution. This suggests that in the immediate proximity of a contaminated source, Archaea are the primary degrading constituent of organic contaminants. Therefore, a thorough examination of Archaea, which produces methane, offers a marker for groundwater contamination. The analytical relevance of the DNA microarray is restricted because of the choice of specific microbial species and PCR settings for the amplification of 16S rRNA genes. Despite these drawbacks and the need to determine microbial concentration, family-specific DNA microarray technology makes it easier to assess the natural attenuation of contaminants along the downstream flow from abandoned landfill sites by providing the composition of microbial communities in groundwater. The site-independent investigation of microbial populations engaged in natural attenuation can be facilitated by the use of bacteria-and archaea-specific DNA microarrays, which are a flexible and efficient method for routine groundwater quality evaluation. Many impact assessments are conducted by taking into account just one or a small number of parameters. The impact on groundwater is assessed using four variables: electrical conductivity, sulfates, chlorides, and nitrates. Researchers used a criterion containing total organic carbon, nutrients, and pesticides to assess the impact of agricultural activities. Nutrient concentrations alone are used by some researchers to evaluate impact. Scientists assess the chemical status of groundwater bodies using a variety of physicochemical factors, nitrogen and phosphorus compounds, and water chemistry. Rather than using data from monitoring networks to determine the impacts, we can use mathematical models. A method for determining if metals have an impact on the chemical condition of water was also proposed.

7.7. HEAVY METALS

A large number of hazardous heavy metals have been released into the environment as industrial effluents, seriously contaminating soil and water. Diverse technologies are needed to remove heavy metals from the environment due to increasingly stringent restrictions. The conventional method of treating heavy metal-contaminated groundwater was pump and treat, which was neither economical nor environmentally friendly. Permeable reactive barriers (PRBs) appear to offer a viable and efficient substitute for the in-situ treatment of groundwater contaminated with heavy metals (Obiri-Nyarko et al., 2014). These barriers can be constructed vertically or horizontally, which allows for penetration into the unsaturated zone and perpendicular groundwater flow direction. Precipitation, ion-exchange, membrane filtration, and sorption are the methods most frequently employed to remove hazardous ions or lower their concentrations in the aqueous phase. Reactive barrier materials shouldn't produce harmful intermediate products, or compounds that negatively react with pollutants. Similarly, the material must be inexpensive, long-lasting, and able to maintain groundwater flow. For the elimination of metal ions from aqueous solutions, such as Zn (II), Cd (II), Cr (VI), Cr (III), Ni (II), and Fe (II), liquid membranes (LMs) and, more recently, polymer inclusion membranes (PIMs) can be an interesting substitute. Transport of various metal species through PIM can be explained as simultaneous extraction and re-extraction processes taking place in a single step. In order to create a thin, flexible, and stable film, PIMs are created by casting a solution that includes an ion carrier, a plasticizer, and a base polymer, such as polyvinyl chloride (PVC) or cellulose triacetate (CTA) (Maiphetlho et al., 2021). Source and receiving phases are separated using the resultant membrane. PIMs are very stable and adaptable, yet they still have most of the benefits of LMs. Furthermore, PIM-based systems can benefit from a number of factors, including ease of use, minimal use of hazardous chemicals, and flexibility in membrane composition to achieve the required selectivity and separation efficiency. This is because PIM mechanical properties are quite similar to those of filtration membranes.

7.8. PIM

PIM technology, which consists of tubular modules made of immobilized membranes, can be used to pre-treat landfill leachates to reduce the concentration of heavy metals in subsurface soil and/or minimize the risk of groundwater pollution. PIMs make it easier for the ion exchange process to eliminate metal ions from aqueous solutions. Ion exchange membranes are a very promising alternative material for the removal of metals because they may avoid the internal diffusion limits produced by resins employed in flow-by mode and are not compressible like ion-exchange resins. The effectiveness of PIMs technology as a separation technique within a PRB for the removal of Zn (II) and Cd (II) from groundwater was investigated. It was also investigated how process variables including the pH of the source phase, the volume ratio of the source to receiving phase, and the molar concentration of DNNS in PIMs affected the counter-ion transport. A substantial source of drinking water is the spring water found in alpine karst aquifers. An automated microbiological sampling instrument and portable hydro-meteorological measurement stations communicated using satellites in Low-Earth-Orbit (LEO) can be used. This made it possible for the data to be distributed online and for registered users to download it. It was shown that after developing particular spring and event-type calibrations that account for the unpredictability of the occurrence and the transferability of fecal material. On-site investigations of the spring water isotopic composition were done with high temporal precision and in sync with other hydrological indicators in order to maximize the potential of these early warning systems. For on-site use, the Cavity Ring Down Spectroscopy (CRDS) was modified.

7.9. WATER SUPPLY

Many European nations rely heavily on the water resources found in alpine and other mountainous karst aquifers for their water supply. The Water Framework Directive governs the sustainable protection of karstic catchments. Establishing early warning systems and quality assurance networks for water supplies is crucial due to the growing influence in these areas and the varying uses in the watersheds of karst springs. These systems mostly rely on in-situ measurements to provide data that is available online almost instantly. It is possible to conduct precipitation-triggered event monitoring campaigns at various karst springs using a satellite-based network of measuring and sampling stations. This is achieved by combining online hydrological parameter measurements with field-laboratory-based analyzes of microbial fecal indicators. We need to investigate the chemical parameters, environmental isotopes, and microbial fecal pollution indicators at a high-resolution time scale during hydrological events. Adding automated precipitation sampling to the catchment area to perform isotopic analyzes can be another important strategy. The spring catchment region, where the event sampling will take place, is where precipitation station (PS) is located. It has a LEO-Satellite modem, a data logger, and a tipping bucket attached. Additional meteorological sensors and sample tools can be added to it. An additional data logger, a pressure probe to record discharge variations, two automatic sampling units (reference sample and periodic samples), and a LEO-Satellite modem for real-time control and data transmission are all installed at the spring monitoring and sampling site. Additional meteorological or hydrological sensors can be added to it. The Central Monitoring Station (CMS) is the network central node. Here, all data is gathered, the technical support team is informed, and the database is updated.

7.10. OBSERVATIONS

Rainfall and other meteorological information are recorded at PS. The PS initiates one or more Spring Sampling Station (SSS) via satellite to collect the reference sample if trigger-level is exceeded (Stadler et al., 2011). This occurs prior to the occurrence having an impact on the spring discharge. Additionally, the CMS receives frequent data sets from the PS via satellite in order to monitor the continued trend of precipitation. Spring Sampling Station (SSS) has an automatic sampling device that collects the reference sample as soon as it receives the activation data-set. Depending on how many sampling bottles are in the automatic sampling system, this process may be done multiple times. The second trigger event, an increase in the discharge, is what the SSS watches for within a fixed amount of time. The trigger level is determined by measuring the gauge height increase over time and taking into account the spring properties. The data logger initiates this trigger requirement. The periodic sampling begins automatically if the predetermined trigger level is surpassed, and the CMS receives continuous satellite data streams containing the measured values and status information. The Central Monitoring Station (CMS) collects data from each and every station. Furthermore, the CMS automatically notifies the local service team (LST) of significant occurrences, such as the beginning of a precipitation event (first trigger) and the commencement of the sample process at the SSS (second trigger), through the Global System for Mobile Communication (GSM). The LST can schedule the next visit to the SSS in order to maintain the station based on the sampling time increment and the quantity of bottles in the automatic samplers. The online portal offered by the CMS allows users to view these environmental data. It is based on the dependable operating system, which offers flawless server performance and interaction. The server has a Redundant Array of Independent Disks (RAID) system for fault-tolerant operation among other things. After automatic sampling, the rainwater was kept for 16 hours in an airtight container before being treated. In a field laboratory, E. Coli was directly analyzed. At intervals of fifteen minutes, online measurements of hydrological parameters were obtained in situ.

The indicator of choice for the research of microbiological fecal contamination was E. Coli. As a general fecal pollution indicator in alpine karst environments, it is more applicable than other standard indicators, as evidenced by several previous studies. Examples of this include high prevalence and abundance in human, livestock, and wildlife excreta, low or no prevalence in alpine soils, and a half-life time of E. coli in spring water within the range of the average event length. The recently developed highly sensitive gas analyzer, the Cavity Ring Down Spectroscopy (CRDS) device, was modified and employed for on-site investigation of the isotopic composition of the spring water (Munksgaard & Nelson, 2021). Almost all minute gas-phase molecules have a distinct near-infrared absorption spectrum, including CO_2, H_2O, H_2S, and NH_3. This is made up of a sequence of narrow lines at sub-atmospheric pressure, each with a distinct wavelength. The concentration of any species can be ascertained by measuring the strength of this absorption, or the height of a particular absorption peak. However, trace gases offer less absorption for measurements in standard infrared spectrometers, usually restricting sensitivity to parts per million at most. By employing an effective path length, CRDS circumvents this sensitivity restriction. It makes it possible to monitor gases at the parts per billion and, in certain cases, parts per trillion level in less than a second. A single-frequency laser diode beam enters a cavity in CRDS that is defined by two or more high reflectivity mirrors. Picarro analyzers employ a three-mirror cavity to sustain a continuous traveling light wave. This offers a better signal-to-noise ratio than a cavity with two mirrors that sustains a standing wave. The cavity rapidly fills with circulating laser light when the laser is turned on. A quick photo-detector generates a signal that is exactly proportionate to the cavity's intensity by sensing the tiny amount of light that leaks through one of the mirrors. Moreover, the final concentration values are especially reliable since they are not affected by variations in laser power or intensity' rather, they are based on the difference in ring down durations.

A second loss mechanism (absorption) within the cavity is created if a gas species that absorbs the laser light is introduced. In comparison to a cavity with no additional absorption because of a targeted gas species, this shortens the ring down time. The ring down time of the cavity with and without absorption caused by the target gas species is automatically and continuously calculated and compared by Picarro instruments. This yields accurate, quantitative readings that account for any variations in intra-cavity losses over time. The caves and hypogenic karsts can be seen as the groundwater flowing within the framework of the regional flow system. In order to characterize the mechanisms operating at the Buda Thermal Karst discharge zone today, a thorough hydrogeological study should be conducted. Mineralogical and hydrogeochemical analyzes were among the techniques that can be used. The findings reveal a number of mechanisms that may be involved in the formation of minerals and the development of caves. Since groundwater is essential to the development of karsts, karst phenomena such as caves and precipitates can be understood as geological expressions of groundwater flow. Based on their relative positions within the groundwater flow system, the epigenic and hypogenic karsts are the two main kinds of continental karst systems. The epigenic karst systems are primarily local systems, or portions of intermediate and regional system recharge segments. Areas of hypogenic karst are linked to regional or intermediate flow network discharge regimes. The Buda Thermal Karst functions as a discharge zone for the regional fluid flow because it is a marginal location at the edge of a sedimentary basin and raised carbonate boundaries (Havril et al., 2016). This suggests that a variety of discharge features, including springs, caverns, and mineral precipitates, could result from it receiving fluid components from many sources. Palaeo phenomena in the Buda Thermal Karst and other hypogenic karst locations with comparable settings can be better identified and understood by analyzing the symptoms of recently active processes in this hypogenic karst area. In order to do this, thorough hydrogeological research was conducted to characterize the mechanisms operating in the Buda Thermal Karst discharge zone. The Rózsadomb and Gellért Hill regions of the discharge zone were selected as popular. These regions experienced the discharge of fluids with different temperatures and chemical compositions, as well as the occurrence of ancient and newly emerging hypogenic caves. Furthermore, the upwelling hydrothermal fluids were previously characterized using deep wells on the Pest side. Researchers state that all waters under investigation are part of the Ca + Mg, Na + K cation facies. There are variations in the anion facies, though: the $Cl + SO_4$, HCO_3 anion facies is found in the wells in South Budapest, the waters surrounding Gellért Hill, and a few deep wells from the (northern) Pest side' the HCO_3, $Cl + SO_4$ facies is found in many wells.

7.11. ARSENIC

The presence of arsenic or As in groundwater poses a significant threat to the environmental health of millions of people globally. Both natural and man-made sources of As are found across the planet, and they have an impact on many areas at both the local and regional levels. The occurrence of increased amounts of arsenic in groundwater resources used as drinking water supplies has been a subject of intense global research due to its hazardous effects on human health. Accordingly, case studies on the detection of naturally occurring arsenic levels in groundwater, the occurrence and distribution mechanisms, the effects on human health, and treatment methods have been documented for Bangladesh, India, Nepal, El Salvador, Ecuador, Honduras, Mexico, Chile, China, Canada, Argentina, Peru, Taiwan, United States, Bolivia, and Turkey. Arsenic can be found naturally in many of these places in the underlying strata of sedimentary and volcanic formations, as well as in regions of geothermal systems connected to tectonic activity. It is also known that large amounts of arsenic in surface and subsurface fluids can arise from the discharge of geothermal waters. Arsenic concentrations in Yellowstone National Park geothermal springs are more than 1 mg/l and as high as 0.36 mg/l in the Madison River downstream from the springs (Thompson, 1979). It is projected that more than 100,000 kg of arsenic derived from geothermal processes escape the western edge of the park annually, impacting the quality of water in a vast area. There have been reports of as high as 27 mg/l of arsenic in the El Tatio geothermal field (Figure 7.2) in North Chile.

Figure 7.2. *El Tatio geothermal field.*

Source: https://en.wikipedia.org/wiki/File:G%C3%A9iseres_del_Tatio,_Atacama,_Chile,_2016–02–01,_DD_01–02_HDR.JPG, Creative Commons License.

Western Anatolia (Turkey) is a region with complicated geology, active tectonics, and significant geothermal potential. The high concentrations of arsenic in underground waters are appropriate for this natural habitat. The recurrent opening and closure of the Palaeozoic and Mesozoic oceans has dominated its geological and tectonic evolution. It is a part of the Mediterranean Earthquake Belt, which was created by the collision of the African and Eurasian plates, causing complicated deformation. These plates are separated by seismic belts that are characterized by active faults and young volcanoes, the latter of which permit both heat and water movement. The distribution of fault systems, volcanism, and hydrothermally altered places in Turkey closely corresponds with the distribution of hot springs. The quality of water supplies may be impacted by changed rock types as a result of volcanism and neo-tectonic structure. Thus, groundwater-laden arsenic is usually of geogenic origin and is closely associated with the regional geology of the area. Specifically, arsenic serves as an indicator for Western Anatolia hot water deposits. It is present in practically all geothermal waters and is employed as a tracer to identify pollution in surface and subsurface waters containing geothermal fluids, along with lithium and boron. The Anatolian Plate has different altered rock types with high amounts of arsenic and other trace elements because of its neo-tectonic structure and the impact of volcanism. Because of the way water and rock interact in geological formations, these rocks show a great potential to affect the quality of water resources. Thus, Western Anatolia has high geogenic arsenic levels observed throughout a large geographic area. In certain areas of this region, values in the milligrams per liter range are typical. In the Algarve, in southern Portugal, surface water from reservoirs supplanted groundwater as the primary source of public water supply until the end of the 20th century. The severe drought that struck the area in 2004 and 2005 highlighted the necessity for an integrated water resource management (IWRM) program and exposed the issues with a single source water supply approach (Fulazzaky, 2014). In the near future, the coordinated use and management of several water sources for various water-consuming activities will be crucial as part of the more intricate idea of integrated water resource management. Surface water, groundwater, and non-traditional supplies like treated wastewater for irrigation are some of these many water sources. In those aquifers that are available for public water supply, some works aim to address the regional quantification of groundwater availability and exploitation sustainability, as well as their dependence on factors like the spatial and temporal distribution of recharge, aquifer heterogeneity, and the location of the pumping wells.

CHAPTER 8

GROUNDWATER TREATMENT SYSTEM

Contents

8.1. INTRODUCTION

Stopping or reducing the amount of waste that enters a body of water such as a river, lake, or ocean is the primary objective of wastewater treatment. The water body cleans itself after we cease adding pollution to it. We treat water to eliminate materials that are incompatible with its intended use, whether it is drinking water or water used for industrial purposes.

We are concerned with the state of the water. Groundwater treatment is distinct, though. We are purifying the water body by treating the groundwater. Just like in other water treatment procedures, we start by eliminating any potential sources of pollution. After that, we have to keep cleaning up the groundwater body. Upon removal of the contaminating sources, the groundwater purifies itself. Aquifers are distinct bodies of water as well. Accessing groundwater is exceedingly challenging due to its confinement and extremely slow flow rate. The uneven distribution of groundwater flow adds to the complexity. Every section of the aquifer has the potential to be contaminated, but some flow faster than others. The combination of these distinctive features of the aquifer results in special cleaning needs. The secondary design challenge is that the cleaning process will alter the aquifer conditions, even if the accessibility issue can be resolved. This implies that the design engineer will face the task of creating a system that can adapt to changing parameters as the project progresses because the cleanup design may change over time, especially for pump-and-treat systems. All groundwater treatment systems must be developed using life cycle characteristics' this design approach is known as life cycle design (Bayer & Finkel, 2006).

8.2. GROUNDWATER REMEDIATION

The next conundrum in groundwater remediation is the fundamental decision of whether to treat the toxins in situ, or in place, or try to remove them from the groundwater. There are two main ways to remove pollutants from groundwater: using water or air as transporters. In order to try to transfer the contaminants to the carrier, we must move the air or water past the contaminated location in both situations. The pollutants can subsequently be transported above ground by the carrier and treated in a system for elimination or destruction.

The pollutants are treated on-site by the alternative principal groundwater treatment technique. Natural chemical and biological reactions, accelerated natural reactions, and physical or chemical immobilization techniques can all be used to achieve this. The equipment required for delivering the necessary reactants or agents with in situ remediation systems must be considered. The inefficiency of air and water in eliminating all pollutants from an aquifer is a significant drawback of groundwater treatment utilizing carriers. Water and air struggle to remove the remaining toxins from the aquifer for hydrogeological and geological reasons that we need to address. This indicates that while a pump-and-treat system performs well at the start of the operation, it is unable to complete the cleanup in a reasonable amount of time. Notwithstanding these restrictions on aboveground treatment systems, researchers have discovered many scenarios where the usage of air and water is necessary as part of groundwater restoration efforts. We must use life cycle strategies to maintain a sustainable treatment system for the duration of the project in order to make economical and efficient use of aboveground treatment equipment.

Identifying the goals of the aboveground treatment system is a prerequisite before choosing specific treatment techniques. To achieve this, we need the mass balance for the entire treatment system. Before selecting the optimal technology to carry out the necessary eliminations/removals, we must comprehend what enters the treatment system (influent) and what the effluent requirements are. To complete our mass balance around the treatment system, we need to include additional potential components after choosing the actual treatment technology. For instance, waste sludge would be produced by heavy metal precipitation treatment techniques and air emissions would come from air strippers. We need to incorporate a thorough mass balance around the treatment system in order to complete our technological and financial analysis. Many of the methods that are employed for treating groundwater were first prepared for treating wastewater from cities and industries.

Even if this equipment is still affordable for treating groundwater, we still need to comprehend and create design standards for aboveground treatment equipment. We should know the differences between design criteria and wastewater treatment specifications as well as how to gather the information needed to create a groundwater treatment design. Flow is a good place to start when defining the parameters that will determine the treatment method and final system design. Flow is a reasonably easy characteristic to measure in water and wastewater treatment. The majority of municipal systems are made to last for 20 years. The engineer calculates the time by multiplying population estimates by a standard/constant factor to account for per capita water demand. An engineer includes flows from the city industrial waste as well as a flow factor for storm sewers and rainwater inflow, depending on how old the sewer system is. The projected flows from each unit activity at the facility are added up to determine the industrial wastewater flow. A storm-water runoff factor is included based on the age of the facility. A safety factor of roughly 20% is added in both situations. In each of these scenarios, flow is a fixed parameter that the engineer must ascertain' it is not a design variable.

8.3. PUMPING SYSTEM

The primary goal of the groundwater pumping system is to collect a predetermined volume of water under one situation. This scenario is employed when we have a fixed requirement for a specified amount of water (municipal drinking water), or when we are seeking to stop a plume from traveling past a set area (catch). The second requirement is when we attempt to use pumping as a component of the aquifer restoration process. During groundwater remediation, flow is a design variable. The many components of the overall flow are not added up by the engineer. Rather, he/she needs to consider how flow affects the overall system cost as well as the amount of time required for a final cleanup. We can accelerate the process of capturing and eliminating the toxins by increasing the water flow through the contaminated part of the aquifer. Although the extra flow shortens the time it takes to attain cleanup, the treatment system's size and expense increase. When constructing a remediation system, it is important to take into account the various variables that affect the flow of groundwater systems. These factors include the kind of rock or soil (aquifer) that the water must move through, the water hydraulic conductivity, its capacity to move through materials, the kind of contaminants that moves through the soil, and the design area hydraulic gradient. There are two parts to subsurface pollutant flow: a vertical part and a horizontal part. After passing through the unsaturated zone, where there is no water, the contaminant comes into contact with the aquifer. Groundwater moves constantly under most conditions, but it normally moves slowly, typically less than a few hundred feet annually. Baseline data is required in order to ascertain the flow and direction within an aquifer. It is possible to compute groundwater flow rate once we have gathered or estimated that data. An aquifer can be penetrated by digging a trench or a vertical or horizontal well to transfer groundwater up to the treatment system. The polluted water is raised to the surface by a pump. Groundwater flow is determined by head differences recorded in the aquifer. Water levels and head relationships alter when we pump water out of the aquifer. It is evident that the total groundwater flow patterns in a groundwater remediation system become extremely complex when just one or two pumping wells are added. This is one of the reasons why groundwater remediation system design requires the help of hydrogeologists.

8.4. ZONE OF INFLUENCE

The zone of influence is the area of drawdown surrounding a pumping well (Méité et al., 2023). The zone of influence and the area around a pumping well must be defined and understood in order for many groundwater remediation technologies to be successful. A correctly placed and designed well in the right hydrogeologic unit has the potential to halt and even reverse the contaminated plume. Contaminated groundwater will be drawn out of the well by the pump and sent to the aboveground treatment system via transportation pipelines. The flow required to halt and/or reverse the migration of the contaminated plume is the primary consideration in the design of the groundwater recovery system. Other techniques for regulating the plume movement include the placement of physical and hydraulic impediments. Several injection wells are usually positioned to change the groundwater flow patterns in hydraulic barriers. To improve the efficiency of a groundwater recovery system, the hydraulic barriers might be positioned downstream of the contamination or at other key spots. Hydraulic barriers have the major drawback of perhaps requiring a large supply of clean water. The actual physical characteristics of the rock or soil, as well as local, state, and federal permitting authorities, may place restrictions on the installation of injection wells. Often, the hydraulic barrier is formed by treated groundwater effluent. Physical barriers can be positioned upstream, downstream, or entirely around the contaminated plume. A physical barrier can separate the contamination from the groundwater flow system when it is properly designed and built. The spread of the plume can be stopped by halting groundwater migration through the contaminated area. Ensuring the overall effectiveness of the subsurface barriers requires defining their physical location. To ascertain whether physical barriers are appropriate, project management by hydrogeologists, engineers experienced with subterranean constructions, and geotechnical engineers is required. The costs and features of the location may place significant restrictions on barrier use. To stop contaminants from moving (horizontally and vertically), these physical barriers must be placed into confining layers' the deeper the confining layer, the more expensive construction will be.

8.5. BARRIERS

Building a physical barrier with enough integrity at depths higher than 60 to 100 feet is challenging and might necessitate the installation of a sophisticated monitoring system to assess the barrier efficacy. Subsurface barriers may not be able to be used due to site features alone. In any case, the price of installing them needs to be weighed against the price of using groundwater recovery to halt the plume flow. Physical barriers, even in the best of circumstances, will only be able to impede the contamination growth. The permeabilities of the geologic units beneath the aquifer base are frequently high enough to permit groundwater exchange between the aquifer and the units below. Furthermore, the aquifer may be thick and a plume may not reach the base of the aquifer. The main premise is that groundwater flow from a trench system or recovery well is one of the elements that regulate the pollution plume, and that the plume needs to be prevented. The design engineer will always need to collaborate with subsurface hydraulics experts and vice versa. The amount of water entering the contaminated area on a local or regional scale is an important aspect that may need to be taken into account. Barrier walls like slurry walls have the ability to stop the flow of materials into or out of the contaminated area. Recharge from rainwater that can seep into the aquifer and recharge from surface water from lakes and streams are two more potential entrance points into the contaminated area. Furthermore, groundwater may be obtained from the bedrock and other sediments beneath the aquifer through upward leakage. Although uncommon, this needs to be taken into account, particularly when extracting water from the upper aquifer.

One option to manage surface runoff is to cover the contaminated area with an impermeable layer or create effective surface area drainage that diverts water away from the region. Because it is more difficult to manage, water from below will most likely need to be added to the treatment system flow. The duration of the cleanup is another component that affects flow to the treatment system. The level of water surrounding a well in most groundwater systems decreases as more water is pumped out of the hole/well. As a result, the groundwater head differential will rise, forcing water into the well more quickly. An aquifer can only hold so much water at a time, and in many cases, pumping a well empty is quite simple. Pumping rate, system architecture, and the amount of time required to clean up a place are frequently related in complex ways (Jones et al., 2009). The expense of expanding the treatment system size vs the savings from speeding up groundwater remediation must be considered by the engineer or project manager. System flow and cleanup time have a complex relationship. Finding the source or sources of pollution is essential to the assessment of any groundwater contamination site. Preventing such sources from adding to the pollution plume that is already growing is a crucial stage. Pollution sources could be subterranean storage tanks, surface sources, or even unused product or pollution in the saturated or unsaturated subsurface. If the source is not found and eliminated, there may be an issue because the unsaturated zone still contains toxins. Three months later, another pollution plume can arise, and these pollutants might still travel to the aquifer. One way to solve this issue is to slow down the initial cleanup process and allow natural flushing mechanisms to remove the majority of the toxins. Naturally, another option would be to maintain the initial cleaning site while accelerating the natural flushing action, or to treat both areas simultaneously by combining a groundwater approach with a vadose zone treatment technique like vapor extraction.

8.6. SEASONAL VARIATIONS

Seasonal variations in the groundwater level are another example, albeit one that is more difficult to resolve. Naturally, there can be an increase in the aquifer overall level during periods of heavy rainfall. When the level drops and the pollutants remain in the unsaturated zone, a problem occurs. The pollutants will return to the aquifer the next time there is a high groundwater level or when rain seeps through the soil. In these situations, the best course of action might be to apply a vadose zone treatment method once again, excavate and remove the contaminated soil, or slow down the flow to the treatment system and spread out the cleanup across multiple high-water seasons. The impact of other factors on the flow to the treatment system will be negligible. The volume of flow required for cleanup will depend on how many recovery wells or trenches are employed. The position of the recovery system determines how good well trenches and wells draw in pollutants and clean up the site. Systems near the borders of the plume might be effective in removing pollutants from the area, but they also need to be built to collect the least quantity of pure water possible from outside the plume. Locations for successful recovery systems must also account for nonflow zones brought on by the system design. Pumps can be turned on and off in order to alternate the pumping schedule and effectively capture nonflow zones. While cleaning up the groundwater is the main goal, there are a lot of other considerations that may be made when choosing where to put recovery systems. For example, it can also be necessary to install a second recovery system close to the contamination core. The likelihood that we will require an additional recovery system or systems to catch the entire plume increases with its size. A further instance is positioning the recovery system within the plume, engulfing the majority of the plume, and permitting the leading edge of the plume to naturally diminish. The pumping system makes sure that no additional pollutants leave the site and that the amount of material that does leave is sufficiently low for the remaining portion of the plume to be eliminated by natural attenuation techniques. It is necessary to take discharge needs into account when designing the flow rate of an aboveground treatment system. If part of the water is to be utilized to flush the unsaturated zone of contaminants, then more flow will be needed. By driving groundwater to the primary recovery point more quickly and creating a greater hydraulic head, water restored to the site can shorten the cleanup period. The hydraulic characteristics of the soils above the groundwater may place restrictions on this. The flow to the treatment system will rise in proportion to the amount of water that enters the site.

8.7. TREATMENT SYSTEM

Throughout the project duration, the quantity of water released from the treatment system is likely to stay constant. Throughout the cleanup, different amounts of water may be reused on-site. This would apply if the water was being utilized for in situ treatment, to flush contaminants from the unsaturated zone, or to raise the hydraulic head at the end of the plume. The project may alter the flow to the treatment system for these scenarios. When the end of the plume is relatively far from the central well or wells, or when more time for cleanup is not available, water may be recycled to raise the hydraulic head and remediate the plume. As the cleanup phase draws to a close, the plume will recede to the initial location of contamination. The use of recycled treated water to aid in the movement of the plume will no longer be necessary. Nevertheless, the project may potentially be expedited by using injection water. Reusing water to remove impurities from the unsaturated zone may have the same effect. By the time the project is almost finished, the toxins will have been removed or reduced, and aquifer cleanup might be all that's needed. The engineer may choose at this stage to skip using groundwater again. The same holds true for in situ treatment. By the time the project is finished, the aquifer contamination content might be low enough to stop the water from being circulated. As with any well-designed water treatment system, flow fluctuation needs to be taken into account. Apart from the aforementioned primary flow considerations, there are additional variables that may impact the flow over the project duration. Drought and power outages are the two main factors that affect groundwater treatment. Groundwater levels naturally fluctuate throughout the year. The majority of designs are predicated on the aquifer maximal water transport capacity. However, water flow may be restricted under low-water conditions, such as droughts. Low flow rates should be accommodated in the design. The design of the system needs to account for power outages or any other mechanical issue that interrupts the flow to the treatment system. These issues can impact system maintenance at any time of year, but they are particularly problematic during the winter.

8.8. CONTAMINATION

A typical method for determining the amounts of pollutants in groundwater is to take a well sample. The materials that were initially lost will determine the kinds of contaminants. Where the well touches the contaminated plume will determine the relative concentration. The initial contamination plume will experience greater dilution, dispersion, and pollutant degradation. It will be necessary to drill multiple wells in order to fully understand the plume (hydrogeology). When choosing which wells to include in the monitoring system, the hydrogeologist might offer some insights. The concentration of the contaminants is affected by a number of other factors besides the distance from the spill site (Abdollahi et al., 2013). These variables include duration, distance to the water table, solubility and density, groundwater flow rate, and amount of contaminants that reach the aquifer. All of these variables will have an impact on the treatment system size and how long it takes to finish the project. There are various components that make up the amount of contamination that reaches the aquifer. The quantity of contamination lost to the soil is an important factor. The first thing to find out is if the contamination source has been found and removed. In certain instances, the ground continues to get contaminated. A landfill is one example of this. Hydrocarbons stuck in the unsaturated zone under a gas station is one example of the several scenarios that may exist. Another illustration of a continuous source of contamination is leachate formation. When water passes through the contaminated zone and continuously adds new contaminants to the groundwater (subsurface), leachate is produced. The hydrogeologist may help in an effort to ascertain the overall amount of material lost to the ground once the source of pollution has been identified and contained. Not all of the debris that falls to the ground/soil will end up in the aquifer. The toxins will be partially absorbed by the unsaturated zone situated above the aquifer. According to many reports, the aquifer typically receives only 50% of the gasoline that spills from a storage tank. The kind of soil in the unsaturated zone, the distance to the aquifer, and the total amount of material spilled all affect this quantity.

The solubility and density of the material spilled determine the component of the pollutant concentration in the aquifer. It is possible that a large percentage of the substances poured into the ground are not water soluble. An insoluble substance does not mix when it reaches the aquifer. The gasoline primary ingredient doesn't get into the aquifer. It extends outward and floats on top of the aquifer. A tiny portion of gasoline does penetrate the aquifer and may generate the characteristic plume. This gasoline is primarily composed of BTEX chemicals, which are generally found in unleaded gasoline. These days, ethyl alcohol and/or methyl tertiary butyl ether (MTBE) are sometimes added to gasoline blends. These two substances are highly soluble. Because gasoline is less dense than water, it typically floats on top of the aquifer. These kinds of compounds are called light nonaqueous phase liquids (LNAPLs). Since they are often lighter than water, straight-chain hydrocarbons will remain on top of the aquifer. Dense nonaqueous phase liquids (DNAPLs) are compounds that are heavier than water and sink to the bottom of the aquifer. The concentration of contaminants in the aquifer will also be impacted by the volume of water flowing through it. The groundwater scientist must determine if it is more cost-effective to restrict groundwater flow using hydraulic controls. These elements come together to create a temporal effect on the concentrations of contaminants. Pollutant concentrations will follow some patterns. The first is the steady concentration that a leachate displays. The contaminants will be replaced by the source as soon as the groundwater system is able to eliminate them if the source of contamination is not eliminated. The concentration won't change unless the pollution source is addressed. However, we can deal with a leachate whenever there is a persistent source of pollution. When the pollution plume is being pulled in the direction of the groundwater removal system is the second pattern that could occur. This mostly affects drinking water from municipal sources.

At first, the well is pure, but as the plume approaches the well, it gets more contaminated. When handling water that will be used for drinking, there are many criteria.

In this instance, the concentration of the contaminants gradually drops if the initial source of contamination is eliminated. Spontaneous chemical and biological reactions, and dilution from nearby groundwater, mass removal is the primary cause of this. Drilling wells and sampling the groundwater within the wells are two methods that can be used to collect concentration data. There are various methods to merge the distinct well data. There are three options available: taking the maximum concentration observed in all the wells, a weighted average accounting for the impact of individual wells, or averaging all the data. The most accurate approach is most likely the weighted average, and the hydrogeologist can help by providing the weighting criteria. Although it is the least reliable method, the maximum concentration is nevertheless employed occasionally. The reason behind this is that the treatment system won't encounter any issues at lower doses if it can allow the maximum concentration. Wastewater is also treated using this technique. However, it is a risky design approach for groundwater. If the concentration falls below a certain point, a number of treatment techniques, heavy metal removal and biological removal, for instance, will fail. Maximum concentration will not be used to optimize other forms of treatment. Precise concentrations are necessary for the appropriate treatment system design. The weighted average is the most effective approach, the more exact the weight factors, the more accurate the final concentration. Most designs overestimate the concentrations entering the treatment system, even with precise concentration measurements, flow estimates, and weighting factors. The issue arises from our need to ensure that all contaminants are effectively removed by the treatment system. Ensuring that the treatment system is capable of handling both the highest anticipated concentration and lesser concentrations while maintaining its flexibility is crucial. The secret to a well-designed groundwater treatment system is flexibility. The ultimate use of the treated water will also influence the design of the treatment plant. The discharge of the water to a surface water body, another treatment system, a direct use, or back into the ground or aquifer is up to the engineer. The size and complexity of the treatment system will be significantly impacted by the discharge requirements for each of these scenarios.

8.9. DISCHARGES

The National Pollutant Discharge Elimination System controls these types of releases. The type of discharge, whether it be direct or indirect, if it falls under a regulated industrial category discharge, and whether bioassay testing is necessary before discharge must all be ascertained. For a direct discharge, the average effluent needs would be as follows: 10 to 30 mg/L of total suspended solids (TSS), 1 to 5 mg/L of oil and grease, < 1 mg/L of ammonia, 1 to 10 mg/L of nitrate, < 1 mg/L of any heavy metal, and between 0.001 and 0.5 mg/L of toxic organics are among the other contaminants that should be avoided. Always inquire about precise discharge requirements from the local regulators. Furthermore, certain standards are unrelated to the contamination at your location. As part of the discharge permit, it is important to test for dioxins at a site. The treatment system may incur considerable additional expenditures as a result of the monitoring and reporting requirements related to authorized discharges. Discharging recovered groundwater to another treatment system, a POTW (Publicly Owned Treatment Works) or an industrial wastewater treatment system, is an additional option for its final disposal (Theodore & Theodore, 2021). The ability to dump groundwater to adjacent sewer lines is one benefit of discharging to the POTW. One benefit of releasing pollutants into the industrial wastewater system is that they can be of a different type and concentration than those typically permitted in an NPDES discharge. Neutralization, heavy metal removal, or pure compound removal may be necessary for the industrial facility. Typically, the POTW stipulates that the amount of water entering the sewer must not surpass the typical level of household waste. Discharge to a POTW may come with substantial expenses. Certain POTWs have a policy of charging a connecting fee on top of the flow-based daily costs. A surcharge may also apply to any chemical whose concentration exceeds the permitted limit. A direct discharge is typically less expensive for high flow systems, but all of these costs add up. When thinking about discharging to a POTW, engineers should make sure they obtain a comprehensive price schedule for their study.

A growing number of incidents involve spills that have impacted sources of drinking water or other uses. In these situations, treating the groundwater is necessary before reusing it. It should be mentioned in these situations that the well would have been abandoned if the pollutant concentrations weren't typically very low. To ascertain the requirements for discharge, it is necessary to evaluate the federal drinking water regulations. It can also be necessary to handle naturally existing substances as part of a drinking water system. Arsenic is an appropriate example. To ensure that the final user finds the discharge acceptable, engineers may need to incorporate these chemicals into the design of their treatment systems. Naturally, the treatment system expenses may increase significantly due to these organic substances. It is always interesting to consider who should pay these additional expenses. The well water may be refilled with groundwater. The recharge device needs to be positioned strategically to influence the plume movement if the water is to be returned to the ground. By doing so, the hydraulic head will rise and the plume might be forced back toward the well. Gravity is the primary factor causing a pure organic pollutant to move when it is released onto or into the ground. In the event that the ground is porous, the spill will descend. The soil porosity will determine how much of the movement spreads laterally. The viscosity of the liquid spilled and the soil porosity will determine how quickly it moves. Before the contaminant reaches the aquifer, it may experience a number of different reactions as it descends. The pollutant may go through any of the following processes: volatilization, adsorption on the soil particles, biological degradation, and, typically to a lesser extent, hydrolysis, oxidation, reduction, and dehydrohalogenation. Furthermore, the contaminant may come into contact with an impermeable barrier (bedrock, clay, etc.) that has the power to halt or modify the downward path.

8.10. SOIL MATRIX

Both the soil matrix and the pollutant affect adsorption on a soil. Adsorption capacity is influenced by the solubility, molecular structure, and octanol–water partition coefficient (K_{ow}) of the contaminant. The distribution of a chemical between the water and octanol phases in contact with one another under equilibrium conditions is represented by the octanol–water partition coefficient. The greater the coefficient of octanol-water partition, the greater the probability of it adhering to soils. Adsorption is directly impacted by the molecular structure, which is related to the polarity and size of the contaminant. If the contamination enters the aquifer, these elements become even more crucial. In the aquifer, adsorption on soil particles will also take place. For this reason, impurities in an aquifer flow more slowly than they do in water. The soil particles adsorb the pollutants and cause them to slow down. Other significant soil matrix parameters that affect adsorption capacity are surface area and organic content (Kaiser & Guggenberger, 2003). More adsorption capability is provided by clay as it has large surface areas. Increased adsorption capacity results from an increase in the number of possible adsorption sites caused by a higher organic content. A small portion of the contamination will be absorbed by each soil particle as it travels through the soil. Adsorbed material remains in the soil and does not flow with the main flow of the pollutant. The spill will eventually deplete in the soil and stop all primary movement if it does not reach an impermeable layer or aquifer first. The porosity of the soil and the absorbability of the contaminant as indicated by its characteristic maximum residual saturation will determine how much soil is needed to adsorb all of the material released. The pollutant will remain in the soil at or below its maximum residual saturation. The likelihood of an entity contaminating groundwater is decreased when it is adsorbed in the soil. Rainwater carrying the portion of the material soluble in water will carry it through the ground and continue to be a source of contamination. Volatile pollutants have the potential to infiltrate the aquifer via passing through the vadose zone. If the adsorbed material is not treated or removed, it can significantly extend the time needed for a groundwater treatment program. Moreover, if the substance is volatile, it may pose a direct risk. In the soil, volatile organic compounds (VOCs) migrate laterally and upward, penetrating buildings and maybe posing an explosive risk or just giving off bad odors (Schulze et al., 2017). Therefore, as part of the treatment, the organic material in the vadose needs to be handled. The source of contamination must be addressed in the groundwater remediation initiatives. One possible source of contamination in the vadose zone is the adsorbed material.

An entity that has been adsorbed on the soil particles may be hard to locate and remove, which is the problem. Unless there are deposits that impede the movement of free pollutants, there is typically relatively less lateral movement in the unsaturated zone. When building a free product recovery well, the site of the compound release is typically close by, provided that location is known. The compounds in the spill can undergo further changes in addition to those caused by volatilization and absorption in water. These additional changes can come from the biota that naturally exists in the soil, oxidation, reduction, hydrolysis, and dehydrohalogenation. These additional mechanisms or soil microorganisms have the ability to alter an organic compound structure. As a result, the organics released initially may not have been the same as those discovered in the soil. Soil contamination source locations have traditionally been located by research techniques such as discrete soil sample collection or passive soil-gas sampling. The issue with these approaches is that a large number of data points are required in order to accurately depict the size of the possible source area. Nonetheless, there exists a balance between the financial resources allocated for conducting an investigation and the acquisition of statistically significant data. Because of this, the data must be analyzed with the knowledge that the real conditions may change slightly from the data representation, and this must be taken into account when designing the system. The aforementioned investigative techniques, despite their drawbacks, can be successfully combined with other site evaluation strategies. For instance, doing a dynamic soil vapor test is a method that is more frequently employed in site investigation. This method entails setting up several sample locations inside the region and creating a vacuum to collect soil vapors for examination. One can more precisely identify the possible boundaries of the contamination by employing a methodical strategy that includes the extraction and analysis of soil vapors from several sites and the installation of new points depending on the data. Discrete soil sampling can be used to confirm soil contamination levels in a more targeted manner when the possible source regions have been more clearly identified. Apart from the aforementioned soil vapor extraction test, carbon dioxide measurement is an additional technique that has been developed to detect petroleum hydrocarbons in the vadose zone. At a polluted site, soil vapor extraction tests find trace levels of soil vapor spread out over a large area. The vadose zone VOC or volatile organic compounds will be detected by the test. These substances may be a direct consequence of material that has been adsorbed or a sign of contaminated groundwater. This approach only detects volatile organic compounds.

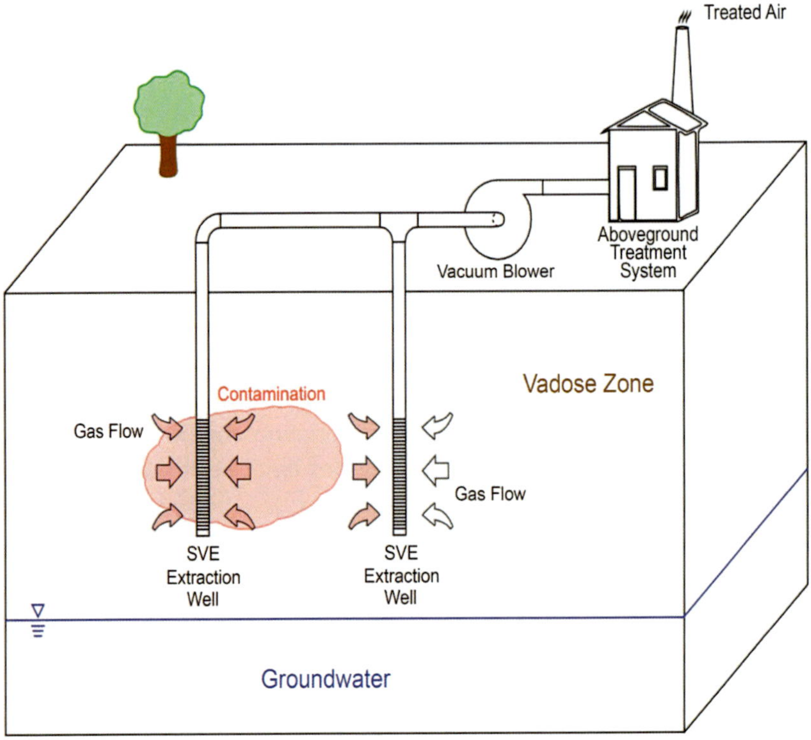

Figure 8.1. *Soil vapor extraction test.*

Source: https://commons.wikimedia.org/wiki/File:Conceptual_Diagram_of_Basic_Soil_Vapor_Extraction_(SVE)_System_for_Vadose_Zone_Remediation.png, Creative Commons License.

8.11. CARBON DIOXIDE

Measuring carbon dioxide is an addition to the soil vapor extraction test (Figure 8.1). Instead of testing for organic vapors, carbon dioxide is detected in the soil vapor for this assay (Baldocchi et al., 1996). Research has demonstrated that elevated carbon dioxide levels occur in the vadose zone as a result of the natural biological breakdown of organic compounds. The only substances that can be used in this approach are biodegradable ones, and organic natural matter may obscure the results. There are various ways to get rid of the substance after it is located. For shallow spills, the ground can be dug up/excavated and either thermally treated, buried in a controlled landfill, or biodegraded in a solid-phase bioreactor like a soil pile. Although flushing and recovery have occasionally been employed, in situ treatment or vapor extraction systems (VES) are the primary techniques of treatment for massive pollution. When an impermeable layer of ground impedes the passage of the pure chemical, many of the same problems arise. Unless the impermeable layer is slant/sloped, the pure compound downward movement ceases when it reaches it, allowing gravity to continue carrying it away from the release location laterally and vertically. The material typically spreads out on top of the impervious layer. This process carries on until the impermeable layer ends and the contamination moves down due to gravity, or until the contaminant comes into contact with enough soil to absorb the spill. When a material comes into contact with an impermeable soil layer, the same problems arise as when a material is adsorbed. It is challenging to collect the material in one location for treatment or recovery. The pollutant can be stopped from spreading further by installing a containment barrier, such as a slurry wall. The slurry wall would need to entirely enclose the spill for it to be effective, and full containment would need to be provided by placing an impermeable cover system atop the slurry wall. It is also possible to dig a well or wells into the impermeable soil and retrieve the pure chemical. Because of interstitial forces in the soil, the well(s) would not be able to collect a large amount of the pure chemical spill. The best cleaning techniques are usually free product removal and in situ treatment or VES.

The substance that spills will either float on top of the aquifer (LNAPL) or sink to the bottom (DNAPL), depending on the relative density of the pure component (Colombano et al., 2020). The contamination concentrations in the groundwater will also be determined by the solubility of the contaminants and the porosity of the aquifer. Two systems, one stationary and the other mobile, where mass transfer happens through molecular diffusion, are commonly used to illustrate soil porosity. That is, there are secondary channels that serve as reservoirs for the contaminants and principal flow pathways that carry most of the fluid movement. This dual-domain model offers a physically grounded mechanism that explains the relief activity and mass storage of relatively immobile solutes in low conductivity materials, in addition to the primary solute transport along small-scale flow channels. The treatment techniques employed for an organic treatment will rely on a number of variables if the aquifer needs to be cleaned. Only in situations when the pollutant is somewhat soluble in water can purely compound recovery be accomplished. The methods for recovering pure compounds will not be effective after the organic material is in solution.

A number of site-specific characteristics, such as the compound composition, hydraulic conductivity, water-table fluctuations, and geology of the site, should be taken into consideration while choosing the optimal technology for pure compound recovery. Important physical characteristics for the petroleum and chlorinated hydrocarbons, including solubility, specific gravity, and partition coefficient should be considered. Pure compound recovery procedures are inapplicable to highly soluble chemicals like phenol and acetone. This also covers the conditions in which an emulsion is produced by a surfactant or other substance. Straight-chain hydrocarbons have a low solubility in water. This includes the majority of petroleum products.

8.12. ORGANICS

Chlorinated organics still have considerable solubility even if they are poorly soluble in water. As a pure chemical, they may have long-term effects on groundwater systems. The chemical is less soluble in groundwater the more chlorine substitutions it has. Many pollutants are still present in the aquifer. Humans do perceive the pollutant presence due to the extremely low threshold level of taste. It has been demonstrated that certain chlorinated hydrocarbons can cause cancer in lab animals. Many of these organics have been the subject of public concern when they were discovered to be present in the potable water system. As a result, these compounds are currently controlled in drinking water systems, with some being reduced to parts per billion (ppb) levels. Once the compounds enter an aquifer, the majority of them will either continue downward until they reach an impermeable soil layer within the aquifer, or until they are bound in the residual phase within the aquifer matrix before reaching this layer. Alternatively, they will continue downward until they encounter an impermeable soil layer within the aquifer, or until they are stopped by an impermeable soil layer above the water table or by the water table itself. The reference compound for the specific gravities is water. As a result, any substance that has a specific gravity of less than or equal to 1 will be lighter than water, and any substance that has a specific gravity of more than or equal to 1.00 will be heavier. Pure petroleum-based compounds are classified as LNAPLs because they are often lighter than water. Since chlorinated compounds are heavier than water, pure chlorinated compounds are classified as DNAPLs. Usually, pure substances/compounds from pipelines, oil terminals, and gas stations can be detected above the water table. To improve the octane rating, gasoline, however, contains a variety of organic ring compounds, the basic structure of which is benzene. While most of the gasoline will float, the soluble components will enter the water due to the relative solubility of these ring compounds in water. The alcohol content of (gasohol) gasolines likewise follows this rule. At the base of an aquifer are typically the pure chemicals from the release of industrial solvents containing compounds like tetrachloroethene and trichloroethene. These compounds, like LNAPLs, leave a trail of residuals from the release site. Tetrachloroethene and trichloroethene will enter groundwater and migrate in the direction of its flow since they are somewhat soluble in water.

The pure component must be confined or removed once it has reached the aquifer in order to prevent it from continuing to pollute the water supply. It is imperative that the remediation system design engineer emphasize the significance of eliminating or confining contaminants in their pure form. LNAPLs are far easier to remove than DNAPLs. For the following reasons, removing a floating material from an aquifer is easier: (a) it is shallower, floating above the water table instead of below it; (b) the aquifer water can be used to direct the floating compound movement; and (c) residuals entrained in the soil matrix above the water table make them accessible to removal mechanisms like soil vapor extraction, whereas residuals for DNAPLs are present both above and below the water table. Although rising water levels in some geologic settings will trap some residuals in the aquifer matrix just below the water table, the LNAPL will fluctuate with changes in the water table. Since DNAPLs become entangled in both the saturated and vadose zones when the DANPL migrates downward within the aquifer, they are exceedingly challenging to extract from an aquifer. DNAPLs will disperse atop the confining units, whereas LNAPLs will disperse atop the groundwater surface. There are other kinds of pumping systems that can be used to remove the LNAPL after it enters the well screen. The cost of the recovery system and the quantity of recovery wells required to remove the LNAPL must be balanced by the design engineer. While it is frequently possible to pump enough groundwater from a single well to decrease groundwater levels sufficiently to restrict the whole spread of the LNAPL, this is not always desired. The LNAPL is being entrained in the aquifer matrix (residuals) as it moves toward the recovery well, which is why it never reaches the well. Installing multiple wells to recover the LNAPL is a method that often reduces the number of residuals entrained and shortens the time needed to remove the LNAPL.

8.13. DNAPL AND LNAPL

Large amounts of groundwater would need to be removed if any method was used on DNAPLs, and there would be minimal improvement in DNAPL recovery (Maire et al., 2018). If recovering DNAPLs is a component of the overall remediation plan for a site, the typical approach is to either install an automated recovery system that prioritizes DNAPL recovery while minimizing groundwater recovery. LNAPL may be removed from the well using a few different techniques when it builds up. First, a bailer is lowered into the well, and the floating material is allowed to overflow the top of the bailer. To ensure that just the pollutant is removed, the bailer lip is kept above the waterline. This method needs a lot of labor and time. This procedure has been mechanized by several manufacturers. Because of their weight, the bailers float in water and sink in materials that are lighter than water. A timer to raise and lower the bailer and a pump to keep pressure in the water table can be found in the same unit. Following the drilling of the well and the setup of the equipment, recurring visits are necessary for both equipment maintenance and recovered fluid management. More specifically, LNAPL and groundwater (total fluids) are usually recovered together. The recovered liquid is subsequently processed with an oil/water separator, with each stream being managed separately, or it is packed for later disposal as a bulk liquid. The amount of LNAPL being recovered will determine whether or not a separation technique is included, and this will also affect how much bulk disposal will cost compared to the cost of buying and operating an oil/water separator. A pump is used by this system to keep the water table stable. The well has an oil/water separator installed. A hydrophobic substance is applied to a separator screen. Water cannot flow through the screen, but petroleum-based products, such as gasoline, can get into the separator. A second pump that extracts the LNAPL from the well is encircled by the screen. Greater material removal rates are achievable with this method.

A skimmer that is powered by pneumatics is intended to float just above the water table and sink in any liquid that is lighter than water. To keep water out of the intake, it is constructed of a hydrophobic substance that floats somewhat above the water table. The operating range of a pneumatic bladder pump is approximately 0 to 5 gpm. Because the float is suspended from a coil of tubing, it can rise and fall in response to changes in the water table. A well with a large diameter is needed for this technique. Two pumps are employed. In this instance, a conductivity probe regulates the LNAPL recovery pump. An electrical charge is carried by water between two electrodes. Electricity is not conducted by LNAPLs. The probe's two electrodes don't transfer any charge while it's submerged in an LNAPL. The second pump is then activated by the unit, which recognizes that it is in a pure compound. In situations where mobilizing LNAPLs toward the well requires hydraulic control and where hydraulic conductivity/permeability is low, it is often more practical and economical to extract groundwater and LNAPL simultaneously and use an oil/water separator to separate the fluids aboveground. The ability to use relatively small-diameter wells in place of the larger wells needed for a dual-pump arrangement of a submersible pump and skimmer pump within the same borehole is one of the main benefits of total fluid extraction. The same benefits that apply to LNAPLs also apply to DNAPLs, where a system with small diameter wells can be employed. The collected fluids would be emptied into a tank with groundwater discharging over the top and DNAPLs accumulating at the bottom. DNAPLs would be removed for short-term storage and disposal as part of a system designed to reduce their buildup at the bottom. Digging a trench in the line of groundwater migration is another method of removing non-flammable LNAPLs. When the spill remains close to the surface, such as in situations with elevated water tables, this method performs best. The LNAPL stays above the water as the groundwater enters the trench. But after the LNAPL is in the trench, skimmers or other common surface treatment tools can remove the floating particles.

The LNAPL movement is regulated by the water in the aquifer. If the LNAPL has the potential to catch fire or poses an explosion risk, then specific designs for this procedure need to be taken into account. For DNAPLs, trenches are typically not employed due to the depths involved. Only in geologic conditions where the DNAPL has spread above a confining layer at a shallow depth may they be employed successfully. DNAPLs might go in the groundwater direction or not. As a result, the compound ability to migrate may or may not be impacted by changes in groundwater flow. The procedure won't be as effective with a floating material, especially when wells can be utilized to push the DNAPLs in a specific direction. The compound may occasionally move against the direction of groundwater flow. Gravity will still have an impact on the DNAPL, which will follow the shape of the aquifer bottom rather than the groundwater flow direction. In order to treat a pure compound discharge that has occurred on the ground, we need to take into account a number of crucial elements. First, we need to ascertain whether the substance has entered the aquifer. The next step is to determine if the material has permeated the aquifer, dissolved into water, or remained on top of it. To recover as much of the material as feasible if it is not in the aquifer, one of the pure compound treatment techniques should be applied. Recovering either LNAPL or DNAPL frequently leads to the recovery of groundwater that has been impacted by dissolved-phase VOCs (Rivett et al., 2011). This water needs to be treated before being disposed of. For many years, dissolved gases including carbon dioxide, hydrogen sulfide, and ammonia have been eliminated from water and wastewater by the use of aeration. It has also been used to oxidize dissolved metals like iron and to provide oxygen to increase the amount of dissolved oxygen. The technology was not used for the treatment of water with low levels of synthetic volatile organic compounds (VOCs) until the mid-1970s.

8.14. AERATION

The contaminated water must be exposed to a fresh air source in order for aeration to occur. The volatile substances in the water are forced out of solution and into the vapor state as the air and water combine. Any aeration system must maximize air/water contact. Towers and aerators are the two main component types utilized in air stripping. Diffused aeration or mechanical aeration are utilized in aeration basins, and packed towers, trayed towers, and spray systems are used in stripping towers. A long array of technologies may be compared before use. For many years, packed-tower aeration was the method most frequently employed to remove volatile organic compounds (VOCs) from water (Linek et al., 1998). However, in the last 20 years, tray strippers have become popular. The target contaminants, air/water flow rates, target removal efficiency, physical constraints, airstream treatment (if necessary), and operation and maintenance factors must all be taken into account when choosing the most suitable technology. Aeration basins are simple, reasonably inexpensive technologies that can be applied in circumstances where volatile organic compounds (VOCs) are present. A continuous-flow tank with air diffusers positioned at the bottom to enhance the gas-phase surface area for the mass transfer of volatile organic compounds is called an aeration basin. The surface area for groundwater-to-air transfer is formed by the bubbles. Dilution causes the mass transfer driving force to be low' however, by adding baffles to the tank, the driving force for the mass transfer can be increased, improving performance. Diffused air or bubble aeration air strippers are typically utilized in conditions with high iron content or other materials that could clog the packed-tower air stripper. A diffused aeration system, which requires less maintenance, is the better option if the water has high iron concentrations. These systems can also be made to fit entirely inside of small buildings at a treatment site, making them portable in design. A few designs additionally incorporate filtration within the unit to provide high-grade air stripper effluent. Because of this flexibility and ease of usage in a variety of low-flow situations, it has been extensively utilized for gas station treatments. To boost the effectiveness of the bubble aeration process, some manufacturers have developed ways to cascade the water being treated through a number of aeration chambers. These systems use a comparatively less amount of area and have a low maintenance.

8.15. MASS TRANSFER

Spray, trayed, and packed towers expand the groundwater's surface area that's accessible for mass transfer. A fan draws air through the tower as the water pours down a tower. Because of the slow flow and the extra surface area of mass transfer provided by cascading water droplets, treatment efficiencies range from moderate to high. Tray aerators come in a variety of forms, all derived from the same stacked tray design. These designs use a spray nozzle to spread water across the upper tray, which subsequently flows into lower trays via the tray sides. Below an orifice tray with a stream of water flowing through it, air is driven in. The purpose of the airflow is to keep the froth in the tray in order to achieve the necessary mass transfer. To improve removal efficiency, a single tray water flow rate can be decreased, or trays can be stacked and run in series to boost system performance. As a result of the absence of packing, fouling is less likely. Above flows of about 100 gpm, these systems are not cost-effective and consume a significant amount of energy. Using a spray basin, stripping tower technology has been applied in multiple groundwater treatment scenarios. These systems spray very small droplets of contaminated water into the air by placing spray nozzles evenly over the area covered by a pipe grid. When it comes to an earthwork recharging basin, the water is collected in the basin and either pumped off or permitted to seep back into the soil or ground. This technique is frequently applied to treat contaminated soil. This system's advantage is its incredibly low capital cost, which makes it perfect for a one-time treatment.

The drawbacks include the potential need for substantial land areas and the possibility of wind-driven mists or, in the winter, ice crystals affecting nearby properties. It could also cost more to pump in order to provide the nozzles enough pressure. Before spraying hazardous material onto land, regulatory considerations must be addressed. A tray tower has two advantages over a traditional packed tower: it is more compact and smaller, and maintenance expenses are reduced because there is no packing involved. This is crucial in situations when there is not much room for the equipment placement and maintenance.

The main drawback is that the tray aerator needs a lot more airflow than a typical packed tower in order to attain a given removal efficiency. Furthermore, many units may be needed to treat the system design flow rate because tray aerators are generally only able to handle lower water flow rates than packed towers. As a result, the tray tower may have higher operating expenses. Therefore, while either kind of system can be employed to meet treatment objectives, site-specific and budgetary considerations play a major role in choosing the best aeration system.

Although packed-tower aeration (Figure 8.2) has historically been the preferred method of air stripping by the government and industry, it is vital to keep in mind that there are other types of air strippers as well. Other aeration designs might be the most practical and cost-effective option for a treatment system in some conditions. A high iron concentration and minimal flows (less than 25 gpm) are two examples of conditions that could result in different designs. One method of mass transfer is air stripping. Air and water are run counter current through a medium in packed-tower aeration systems. By breaking the water into thin films and exposing a large portion of the liquid surface area to the air, the medium improves air/liquid contact.

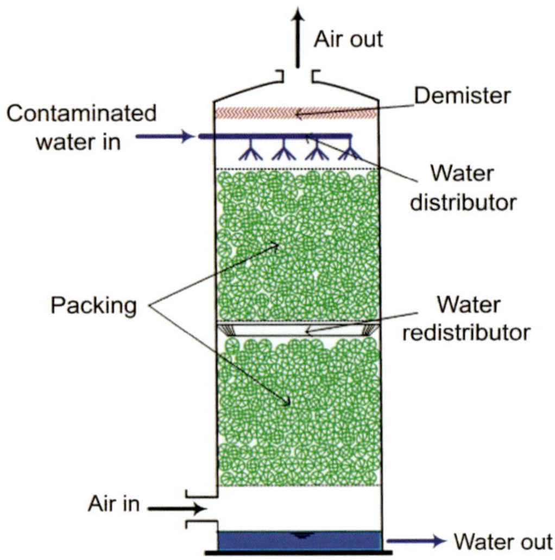

Figure 8.2. *Packed-tower aeration.*

Source: https://commons.wikimedia.org/wiki/File:Air_Stripper_for_Wikipedia.png, Creative Commons License.

There is a greater chance of VOCs from the water transferring into the passing air the more surface area that is exposed (Wolkoff et al., 1998). In order to prevent the stripping process from being hampered by the VOCs' diffusion through the water, the media also act to continuously mix the water. The pollutants from the stripper are carried by the air and either enter the atmosphere or end up in an air treatment system. The treated water passes from the column for use, discharge, or further treatment, if necessary. When the VOCs are discharged directly to air, they are dispersed by the velocity of the airstream and by any wind currents, where the chemicals may break down by natural ultraviolet (UV) degradation or other methods (chemical or biochemical). The usage of airstream treatment technologies may be required in some circumstances due to local legislation, the quantity of volatile organic compounds emitted in the contaminated airstream, or both. The design engineer must specify some fundamental variables when sizing a packed column: (a) Tower height' (b) Tower cross section' and (c) Air-to-Water Ratio.

8.16. TREATMENT

Treatment efficiency is constant for a given bed depth, meaning that if 10 feet of packing removes 90% of the material, another 10 feet will remove 90% of the material left, for a total removal of 99%. The air-to-water ratio depends on the substance that needs to be eliminated. The less air that is needed to remove a compound, the more volatile the substance. Depending on the compound being removed, air-to-water ratios can be from 10:1 to 200:1. The temperature of the water is another crucial factor in the design of air-stripping. Higher temperatures increase constituent instability, which has an impact on the air-stripping process. The temperature of the water that needs to be treated is generally something that the design engineer needs to know. The temperature of the water entering the air stripper can be changed by pretreatment equipment because groundwater varies in temperature. Although air temperatures can vary over a 100°F range, ambient temperature strippers do not have change in working performance from winter to summer operation. This is due to two factors: the thermal mass of the water is far larger than the thermal mass of the air in a stripper, and the groundwater temperature remains at or close to a constant temperature throughout the year. As a result, the stripper's actual working temperature stays rather consistent. Preheaters on the water stream or direct steam injection into the tower can both be used to alter the temperature. Elevated temperatures accelerate the process of stripping, and chemicals that are rarely volatile at room temperature can be completely eliminated by increasing the column temperature to between 140 and 180°F. High-temperature strippers, also known as steam stripping, are often only used for quick treatment of hazardous waste sites or when air treatment is necessary due to high organic content.

8.17. DESIGN

Any packed column design approach begins with the fundamental mass transfer procedure. The driving force (concentration gradient between water and air) and the area of the air/water interface will determine the rate of VOC transfer (Braida & Ong, 1998). Depending on the impurities, different chemicals will transfer at different rates as determined by Henry constant. The ratio of a compound partial pressure in air to its mole fraction in water at equilibrium is known as the Henry constant. In equilibrium, compounds with high Henry constants are found in higher concentrations in air. These compounds are easily stripped because they can transition from a dissolved state to a vapor state easily. Conversely, compounds with low Henry's law constants are more hydrophilic. The temperature has an impact on Henry's law constant. Depending on the necessary removal efficiency, the compounds and concentrations involved, the supporting equipment needed, and the requirement for air treatment, treatment costs for packed towers might vary significantly. The mass transfer efficiency of every process is intimately linked to its capital cost. The process hydraulics and the way the gas is dispersed determine the majority of the operating costs. Since equipment is quite simple, complexity or heavy maintenance is usually not a big concern' nonetheless, it should be recognized that fouling by chemical precipitation and/or biological development needs to be controlled. Aluminum, fiberglass, stainless steel, coated carbon steel, and concrete are among the materials used in construction. The least expensive materials are fiberglass and aluminum, whereas coated carbon steel and stainless steel are often a little more expensive. Concrete towers are the costliest. It is crucial to keep in mind that the cost of raw materials might change quickly, thus while designing a tower, all relevant materials should be taken into account. Aside from cost, the strength and resistance to corrosion of the different materials determine their relative advantages. Aluminum is lightweight, has good structural qualities, and can be used in situations involving potable water. It shouldn't be applied to process streams that are extremely acidic ($pH < 4.5$) or contain high concentrations of heavy metals or chlorides. Fiberglass towers are quite brittle, which makes it challenging to build tall towers subject to wind stresses. However, they offer considerable corrosion resistance in most chemical conditions.

Many of the stainless-steel alloys have outstanding structural qualities and a broad range of corrosion resistance, but they are not easily accessible without paying high prices. Epoxy-coated carbon steel is strong and resistant to corrosion at a reasonable price, although painting and routine internal inspections may increase maintenance expenses. Concrete towers are field-erected structures. Inadequate concrete pouring can also result in honeycombing and leakage in these units. The purpose of the tower internals is to guarantee that the mass transfer process occurs under ideal conditions and at the lowest possible cost. The air exhaust ports are the first part of the tower that needs to be attended to by a design engineer.

These ports are usually found surrounding the tower and are of a size that allows air to exit with the least amount of pressure loss. Towers intended for potable water should have their outlets screened to prevent particles carried by the wind from contaminating the tower. Towers that use a 24-mesh screen have not experienced any issues in this regard. Towers that handle huge volumes of highly contaminated water would need additional ductwork to transfer the off-gas to an air treatment unit or tall stacks to divert exhaust air up and away from the surrounding region. A Mist eliminator system is installed in the tower to stop a lot of water entrained in the airstream from escaping. To do this, the airstream is forced to bend via several points in order to impinge on the water droplets on the mist eliminator's surface. Mist eliminators can be divided into two main types: pads and chevron-style.

The chevron-type eliminators consist of angled plates that are pushed to zigzag through the air, impinging water on the plates. Typically, polyvinyl chloride (PVC), fiberglass-reinforced plastic (FRP), stainless steel, or aluminum are used to manufacture the chevron baffles. The air flows through the baffles in four to twelve different directions on average. The efficiency of eliminating mist is positively correlated with the depth of the baffle design and the centerline distance between baffles.

8.18. FILAMENTS

Filaments are loosely bundled into 3 to 12 inch thick pads to create the pad-type mist eliminators. Water droplets are produced on the filaments, which are often constructed of stainless steel or polypropylene, as air passes through the pad. Although they have a higher pressure drop, pad-type mist eliminators can remove more mist, particularly at higher airflow rates. Furthermore, pad-style mist eliminators are typically less expensive. A significant advancement in mass transfer technology is the use of low-cost plastic packaging. Injection molding has made it possible to create packings that are far better suited to the two objectives of minimum pressure drop and maximum mass transfer surface. All of the polypropylene packings have intricate forms that ensure a large void area to reduce pressure drop. Since it is chemically inert, it won't break down in the presence of most pollutants found in contaminated groundwater. Extreme concentrations of some chlorinated compounds, especially methylene chloride, more than 1,000 parts per million (ppm) might cause the polypropylene to soften and the mass transfer mechanism to deteriorate as a result (Iranpour et al., 2005). The cost of polypropylene is quite low compared to the eight to ten times higher cost of ceramic packaging for an equivalent volume. Because polypropylene is robust and lightweight, it can accommodate deeper packed beds without destroying the packing near the tower base. Ceramic packing needs to be loaded into a tower that is filled with water to soften its fall' nevertheless, due to its strength, the packing can be thrown into a tower without breaking. Installing a tower that is designed to bear such hydraulic loading comes at an additional cost. More durable plastics, like Teflon or Kynan, are available in conditions where polypropylene is inappropriate, but their costs are fourteen to forty times higher than those of polypropylene. In most severe conditions, stainless-steel packings are more affordable than Kynan or Teflon, despite the fact that stainless-steel packings are often less effective than those composed of plastic.

8.19. PACKING

The other major kind of tower packings is called structured packing. These packings are arranged to ensure uniform water redistribution and are physically stacked in the tower. The comparatively low mass transfer efficiency, high capital cost, and increased work required for installation make these packings unsuitable for use in packed-column air stripping. They provide reduced head-loss characteristics, allowing for larger air-to-water ratios at the same power cost. Although better fouling resistance has been claimed, field activities have not supported these claims. The air supply system is the last part of an air-stripping system. Usually, if noise levels are an issue, a forced draft blower is located at the base of the tower or in a building close by. However, a blower size and noise level can be reduced with appropriate tower construction, the choice of a minimal air-to-water ratio, and the reduction of fan speed. If desired, sound mufflers can be installed above the air inlet, but doing so increases the pressure drop and reduces noise levels. Air filters are necessary on some potable water installations to guarantee that no air particles come into contact with the water. An air stripper can be quite easy to operate once it is installed. The water supply pump(s) regulates the air delivery system, turning on the blowers prior to adding water to the tower. The blower is the sole moving component, so maintaining the mechanical system is not too difficult. As a result, little maintenance is needed, just an occasional check of the packing. But occasionally, maintenance can have a significant role in how a tower operates. There are a number of places where maintenance needs might rise. The materials used in construction, the surrounding environment, and the quality of the water will all affect overall system maintenance. For example, the tower internals or carbon steel shell will need to be recoated on a regular basis when carbon steel is used. In order to stop UV deterioration, FRP must also be maintained by recoating any exposed FRP components. It has been noted that snow and leaves can clog air intake structures, and that frozen pipes can result from poorly built systems. By building the tower and its interiors to be completely self-draining and preventing any inside pockets where more than a thin layer of ice could form, freezing can be prevented. Heat-traced, insulated normally-open solenoids for inlet and outlet pipes are required to allow exposed pipe drainage during shutdown. When the air-stripping tower is operating normally, freezing is not an issue unless the inlet water temperature falls below 35°F or in exceptionally cold regions, like Alaska. In these situations, temperature affects Henry's law constant and the rate of mass transfer, which in turn affects tower and equipment performance.

When considering the usage of packed-column aeration systems, there are certain inorganic water quality factors that provide issues with maintenance and operation. The impacts of suspended particles, high microbial populations (degradable organics), hardness, and dissolved iron are particularly important (Naveen et al., 2017). Metals that have dissolved, including iron and manganese, are oxidized during the aeration process. Manganese deposition is usually not an issue because of the pH of the water, but in air-stripping operations, the conversion of ferrous iron to ferric iron can be a serious and sometimes disastrous issue. System efficiency will decrease as a result of the oxidized iron that is deposited on the packing material and eventually builds up to bridge and clog the packed bed. The amount of dissolved iron in the water as well as other elements of water quality like pH and dissolved particles directly influence how much iron deposition impacts system performance. The most underappreciated issue with air strippers in particular and groundwater treatment systems in general is iron deposition. The issue arises from the remedial investigation failure to include iron in the analysis as it is not regarded as a contaminant. To properly build an air stripper, the design engineer needs to know the water concentration of iron, manganese, hardness, and other inorganic elements. Cleaning of the packaging may be necessary as frequently as once a month, depending on the concentration of iron (Fe). Fouling may result in reduced removal efficiency and increased maintenance costs, which may necessitate pretreatment or the use of different air stripper designs (Castro & Zander, 1995). Pretreatment equipment may be just as expensive as the air stripper itself. It is simple to extract the iron from water on its own. By using pretreatment techniques like chemical sequestration and aeration/filtration, iron deposition can be managed. Iron can be easily extracted from water in situations where oxygen is a concern in the tower. The iron will separate from its solution and become insoluble iron hydroxide if the groundwater is aerated before being discharged to the air stripper.

The separation of solids from liquids is the issue when the iron is solid. A filter or clarifier is used to remove iron particles from water. Typically, there are not enough suspended particulates present to require a filter. Groundwater can reach the packed tower once the solids have been removed. More processing is needed for the solids. For ultimate disposal, which will most likely take place in a landfill, the solids must essentially be reduced to the smallest volume feasible. It will be necessary to send the solids via a thickener and then a filter press, belt press, or other dewatering device. It is evident that many unit operations are necessary in order to extract the iron from the water. All of the additional units come with additional construction expenditures, and the cost of operations will go up due to increased labor, chemicals, electricity, and sludge disposal. These expenses will most likely exceed the air stripper on large-scale installations. The design engineer for a small-scale system will likely decide to choose a different air stripper design if the pretreatment expenses are too high. For instance, high iron concentrations can be utilized with diffused aeration strippers without the iron having any negative effects. A biological accumulation within the packed bed may result from high quantities of degradable organics and/or microbiological bacterial populations. The highly oxygenated water and packing material provide an ideal habitat for microbial development, which is why this issue arises. Similar to oxidized iron, a biological accumulation can cause the functionality of the system to decline. In packed-column systems that purify groundwater for use in municipal drinking water applications, biological buildups are comparatively rare. When there are larger (>10 mg/L) quantities of degradable organics, landfill leachate treatment, or groundwater treatment following petroleum spills are more likely to encounter this issue. Chlorine or other oxidizing solutions are typically applied either continuously or intermittently before to the air stripper in order to control this issue (McCarty, 2010). When adding oxidizing agents to densely packed towers, take this precaution: The pace at which the inorganic components of the groundwater oxidize can also be accelerated by the oxidizing agents. For instance, manganese that ordinarily remains in solution may precipitate when a biological deposit is removed by chlorine oxidation.

Using a biological treatment unit as a pretreatment can help reduce the accumulation of microorganisms. The non-degradable organics can be removed by the air stripper after the degradable organics have been removed by the biological unit. Another factor to consider for operations and maintenance is calcium hardness. The loss of dissolved CO_2 from the water during the air-stripping process increases the possibility of the water becoming unstable. Calcium carbonate deposition may result from the removal of carbon dioxide in the packed tower and in any post-treatment distribution system. If there is enough instability and it lasts for a long time, calcium carbonate will clog the packing. The issue can be managed by applying pretreatment or by reducing the air-to-water ratio as much as feasible. Chemical sequestering agents or softening systems can be used as pretreatment. Maintenance issues have caused packed-tower air strippers to malfunction more often than bad design. Periodically, remove the plastic packaging to allow the precipitate to break off. This is a labor-intensive procedure.

Another choice is to acid wash the plastic packaging, although this method is not advised because it severely erodes the plastic and makes it extremely fragile. Because precipitates can form in hard water within weeks, conditioning chemicals may occasionally need to be added to the cleaning process. Furthermore, managing and discarding waste streams (cleaning fluids) will raise project operating costs, which may restrict the use of this technology. The release of VOC into the atmosphere is one of the main issues with air stripping. In a mass transfer/transport method, the contamination is just transferred into a different media' it is not eliminated. The first is the dilution that occurs in the tower prior to the vapors being released, which lessens the consequences of these atmospheric discharges. Commonly used air-to-water ratios range from 25:1 to 250:1. As a result, when the toxin is released into the air, it is diluted by an equivalent amount. In addition to this dilution, as soon as the airstream disperses into the atmosphere, further dilution happens naturally. Another reason is that a lot of substances, including trichloroethene and tetrachloroethene, will decompose in the atmosphere when exposed to solar radiation. For example, the half-life of trichloroethene in the atmosphere is around a day and a half. However, the environmental treatment in the US has focused on destruction or ultimate disposal rather than moving the contamination from one media to another.

The majority of states mandate that stripping tower air emissions undergo treatment prior to atmospheric release. The exhaust gases are often treated using activated carbon, incineration, or chemical destruction in some states and situations where the overall discharge to the atmosphere is too high. Granular activated carbon in the vapor phase is a widely used control method for air strippers. The primary distinction between the adsorption process in the liquid phase and the vapor state is the fluid that is being treated. Adsorption and desorption are the two modes of operation of a typical regenerable VPGAC or vapor-phase granular activated carbon system. The air to be treated is run through the bed of activated carbon during the adsorption phase. There are many variations of bed configurations. The configuration is selected in accordance with the compounds to be treated and the necessary design efficiency. When performing air stripping activities, it is often always necessary to preheat the air. The first thing the design engineer needs to make sure of is that the water in the airstream doesn't evaporate and settle on the carbon granules.

CHAPTER 9

INTRODUCTION TO GROUNDWATER MODELING

Contents

9.1. INTRODUCTION

The relationship between groundwater and surface water has been a simple one-way transport process. Recent in-depth analysis has demonstrated that flow systems can be complex. Intricate flow and mixing patterns can significantly affect the biological, biogeochemical, and physical processes as well as the movement of contaminants. It is necessary to evaluate how these intricate processes affect the risk to the environment and public health. The interaction between surface water and ground water on a spatial and temporal scale was the subject of several investigations in academia and industry. The regulatory structure pertaining to mixing zones was examined, and the policy implications for the interaction between surface and groundwater were considered. According to the U.S. Environmental Protection Agency (USEPA), the first two inches of sediment deposits are where the majority of the exchange activities between the sediment and the surface water take place. Differentiating and describing the many inputs to the surface water sediment system are crucial steps in determining the ground water contribution. Hydrologic and physicochemical approaches were the main methods used until recently to quantify the regional extent and quality of contaminated groundwater discharges and their pollution load to surface waterways. The use of biological indicators, organisms that spend all or a portion of their life cycle in contact with groundwater, to identify zones of groundwater surface – water interaction is a promising area. We need to understand the constraints of numerical modeling as well as the underlying hydrodynamic, chemical, and biological processes to explain the transport of contaminants between surface water and groundwater.

9.2. SURFACE WATER

Surface water bodies are essential components of groundwater flow networks (Sophocleous, 2002). Surface water and groundwater interact in almost every type of environment, from major river basins and coasts to small streams, lakes, and wetlands in headwater zones. The characterization of water transfer between surface water bodies and subsurface water is rather well understood. Streams, for instance, are either increasing or decreasing. On a smaller scale, though, the flow system close to the interface between the sediment bed and the surface water column can be intricate. It is probably easy to think of ground water surface water interactions at this size as a superposition of flows that happen at several distinct geographical scales and frequently vary periodically or in response to a climatic event. Many physical factors and processes, including seasonally high surface water levels, ground water evaporating around the periphery of surface water bodies, tidal fluctuations, limited hydraulic exchange due to impermeable sediment, and streamflow and velocity, can result in complex small-scale flows.

In certain surface water bodies, certain processes may be more significant than in others. For example, the interplay between surface and ground water is extremely unique in streams. Based on the biological environment, a section of the transitional zone and biologically active zone inside streams is referred to as the hyporheic zone. The flow of solutes and water in both directions across a streambed is referred to as hyporheic exchange. Abrupt changes in the bed slope or meanders in the stream channel are frequently linked to the direction of seepage through the streambed. The kind of material in the streambed and banks, the slope and variability of the streambed, and hydraulic gradients all affect the hyporheic zone dimensions. In aquatic systems, the hyporheic zone is a potentially important zone of biological activity. The chemical and biological properties of water in the hyporheic zone might vary greatly from those of nearby surface water and groundwater systems due to mixing between surface and groundwater.

9.3. MIXING ZONES

Although mixing zones have been used for point source discharges, regulatory management has not extensively addressed their use for nonpoint sources. Furthermore, there is typically a difference between the actual physical mixing zone that results from a hydrodynamic sequence of events and the spatial definition of a regulatory mixing zone. The mixing zone is designated as an impact zone where specific water quality standards can be exceeded, provided that acutely hazardous situations are avoided. Hazardous, and bioaccumulative chemicals of concern (BCCs), such as mercury, dichlorodiphenyltrichloroethane (DDT), polychlorinated biphenyls (PCBs), and dioxins, the USEPA has reviewed the possibility of a phase-out of mixing zones. The use of BCCs in authorized discharges to the Great Lakes may gradually end. States have the power to decide when it comes to mixing zones, however the USEPA does offer certain guidelines that might be relevant for nonpoint sources (Hardy & Koontz, 2008). For instance, nonpoint sources are not expressly excluded from the mixing zone definition in the handbook (USEPA. The handbook also states that biological criteria should be used to determine the position and form of the mixing zone, and that the zone can be defined based on volume. The state/region has the authority to decide whether to create a mixing zone and how big it should be. With the exception of states where state regulations forbid them, the USEPA permits the use of mixing zones in some applications. State laws pertaining to rivers and streams typically place restrictions on the cross-sectional area and widths of mixing zones. It is evident from those regulations that nonpoint source compliance was also considered. Regarding lakes, estuaries, and coastal waters, certain states have regulations that define the surface area that the discharge may impact. The underlying water column and benthic region are typically included in the surface area constraint. The precise form and size of the mixing zone are decided case by case when there are no set dimensions. Special mixing zone definitions have been created for the discharge of municipal wastewater into the sea/ocean. According to USEPA regulations, the ZID (Zone of Initial Dilution) must be a regularly shaped area (such as rectangular or circular) that encircles the discharge structure (submerged pipe) and includes the areas where the concentrations of pollutants are high under the design conditions. As long as no other mixing zone regulations are violated, the USEPA will generally accept restricting boundaries defined by dimensions equal to the water depth measured horizontally from any point of the discharge structure. The ZID is frequently indicated differently. The ZID is frequently used to describe the early dilution of a discharge. The process of forcing ambient water into a discharge plume by buoyancy induced turbulence and shear processes is known as initial dilution. Water is drawn into the discharge flow as it spreads into the surrounding water, diluting the discharge. The plume density approaches the ambient density during the entrainment process, and depending on the location at which this occurs, the plume either reaches the surface or gets trapped at an intermediate level. Because first dilution occurs there, the mixing region spatial extent is commonly referred to as the ZID. Ambient mixing processes typically regulate additional plume dilution after the ZID.

9.4. CCC AND CMC

Two water quality guidelines are maintained by the USEPA for the permissible concentration of harmful substances: a criterion continuous concentration (CCC) to prevent chronic impacts and a criterion maximum concentration (CMC) to prevent acute or fatal effects (Yan et al., 2015). The CMC value is typically more restrictive and is more than or equal to the CCC value. At the boundary of the same regulatory mixing zone designated for conventional and nonconventional discharges, the CCC must be satisfied. The hyporheic zone, which can stretch up to about 40 inches (100 cm) below the sediment-water contact, is a zone where surface water transitions from groundwater. The hyporheic zone extent can change with the seasons and in response to drought or flooding. The geologic features of the zone and the dominant hydraulic heads determine the relative contributions of surface water and groundwater to this transition. This zone can indicate a potentially important zone of biological activity in aquatic systems, in addition to being an interesting hydrogeologic feature. It is difficult to define the composition and traits of the biological community that inhabits the hyporheic zone. This is partly because it is generally difficult to get high-quality samples from this zone. As a result, the zone has not received as much attention as other aquatic system components. Investigations have revealed a diverse community of organisms that inhabit substrates at depths greater than 6 in. or 15 cm. Traditionally, scientists have only considered the upper few inches (upper 6 in. or 15 cm) as the biologically active zone in most aquatic systems. The hyporheic zone contains both transient and permanent biological species, some of which only spend a portion of their life cycles there. Permanent species finish their whole life cycle in the zone. Rotifer, oligochaete, copepod, ostracod, crustacean, and other species are among the permanent members. Transient members are species that are normally found as members of the streambed benthos and migrate as early larval stages to avoid disturbances like high river flows or stresses like high temperatures. Within aquatic systems, the hyporheic zone can play a significant role in nutrient recycling in addition to providing habitat for particular types of organisms.

9.5. EVALUATION OF POLLUTANTS

Environmental studies frequently include testing protocols that have been devised to evaluate the possible effects of pollutants on these populations. This is different from our knowledge of the biological community linked to the hyporheic zone and its significance for the operation of aquatic ecosystems. It is a crucial location for nutrient processing and storage and it mediates the availability of N and P. The challenge of acquiring quantifiable biological samples has impeded research on the biological community of the hyporheic zone. Risk assessments are the primary means by which the ecological concerns connected to the release of contaminated ground water into surface water are assessed. Despite the fact that groundwater can be a significant conduit for exposure to aquatic environments, the hazards associated with such releases are usually limited to submerged plants or benthic communities. The hyporheic community, which includes both the microbial and deep-dwelling macrobenthic species that make up the meiofaunal community, is not taken into account in these evaluations. It is unknown how significant it is to leave the hyporheic group out of risk evaluations. It may still be necessary to take into account the possibility of damage to organisms that inhabit the hyporheic zone at some places.

The interaction of the receiving water body conditions (ambient) and the discharge parameters determines the mixing behavior of a point source discharge. The kinetic properties of the receiving water body (stream, river, lake or coastal waters) define the ambient conditions there. The velocity and density distribution in the water body determine the dynamic properties. Since the mixing processes often take a few minutes to an hour, these conditions can often be interpreted as stable states with minimal change. Certain situations, especially those involving tide-influenced flows, may involve extremely unpredictable environmental variables, making it inaccurate to assume steady-state conditions. In this instance, the discharge plume effective dilution may be less than it would be in a steady-state condition. The geometry of a single port discharge is determined by the port's diameter, elevation above the bottom, and orientation' additional geometric features for multiport diffuser installations include the orientation of the diffuser line, the arrangement of the individual ports along it, and surface discharges, the cross section and orientation of the flow entering the ambient water are significant.

The combined effect of the gravitational acceleration and the relative density differential between the ambient conditions and the effluent discharge is represented by the buoyancy. It gauges the effluent flow propensity to rise (positive buoyancy) or fall (negative buoyancy). A sufficient description of the source is necessary for the source characterization of the recharge area in surface water-groundwater interactions. For mixing zone assessments, it is especially important to accurately define the area of the recharge zone, the temperature of the recharge water, the local bathymetry close to the recharge area, the downstream ambient water body cross section, and the surface chemistry of the water-sediment interface.

It is possible to think of the hydrodynamics of an effluent continually discharging into a receiving water body as a mixing process taking place in two different locations. The jet trajectory and mixing in a region are influenced by the source geometry, buoyancy, and discharge momentum. The features of the source become less significant as the turbulent plume moves away from it. The turbulent plume dilution is governed by the ambient environmental conditions, which also cause passive diffusion from ambient turbulence and buoyant motions. To accurately forecast the properties of the mixing zone, it is important to have knowledge about the density distribution of the surrounding water body. Most mixing zone analysis uses ambient density as the governing parameter, not temperature.

The ambient temperature can be utilized to determine the ambient density if the ambient water is fresh and above 4°C. Density needs to be specified explicitly if the surrounding body of water is brackish or below 4°C. For mixing zone analysis, it is practical to ignore vertical variation in density of less than 0.1 kg/m^3 or temperature of less than 1°C. The average temperature or the average ambient density can be specified for uniform conditions. If there is any uncertainty regarding the specification of the ambient density values, it is advisable to start with the fundamental approach. Subsequent mixing zone simulations can be used to investigate the sensitivity of a particular assumption. For instance, the CORMIX family of models, which is used to model a variety of discharge sources in mixing zone analysis, has undergone rigorous peer review by the USEPA and scientific journals.

9.6. MODELS

Regulators (NPDES) for point source discharges use the CORMIX models. In order to evaluate mixing zone features, optimize outfall design, ascertain regulatory compliance, and analyze environmental impact, CORMIX models are employed. The necessity to characterize the sources of contamination is the main barrier to using models for groundwater-surface water interaction mixing zones. Standard source conditions for these discharge sources can be modeled by current CORMIX models. The USEPA and other federal agencies play an important role in the selection and application of simulation models. This should involve establishing standards for appropriate performance and validation metrics and models. Federal agencies should be active when it comes to developing models through sponsored research at the relevant program offices. The U.S. Air Force used a ground water quality modeling advisory system to look into several remediation options for cleaning up subsurface contamination (Suthersan et al., 2016). The system can take uncertainty into account when optimizing the remediation plan through restrictions/barriers as well as when predicting solute transport. Using an approach that has been independently tested using machine learning codes, the system helps users choose the best transport models. Most pollutant fate and transport models available today do not integrate effectively over a variety of space and time scales. In the case of mixing zone models, this is true. Linking near-field effects with far-field effects is a clear and ongoing necessity in mixing zone models' yet, there are currently no standard operating procedures or recommendations for achieving this linkage. Strictly mechanistic models are typically insufficient and occasionally inappropriate to use as the foundation for management decisions, especially when used to replicate intricate processes related to water quality. Studies on partial error propagation for models similar to CE-QUAL-ICM, for example, have consistently produced high prediction errors for variables other than salinity, suspended sediment, hydrologic or hydrodynamic processes (Khangaonkar et al., 2012). These models represent relationships and knowledge about processes and linkages, and they help the analyst organize and gather the data required to make decisions. Natural processes, however, are far more complex than these models often capture.

Regardless of how intricate the process is, all contaminant transport models necessitate making some assumptions about the characteristics of the transport mechanisms. Because of this, they are only able to approximate the risk associated with the actual spread or deposition of a contaminant from a particular site. The analyst and those in charge of making decisions are faced with a challenging condition. Whether directly or indirectly, the analyst must select a transport model by weighing the trade-offs between the higher accuracy of complicated models and the simpler models' easier application and less demanding data requirements. The analysis of contamination risk is likely to have a substantial amount of uncertainty even in the case of the selection of an adequate transport model. For instance, there is frequently sufficient reason to question the correctness of the input data in groundwater solute transport models, which require predicting factors like hydraulic conductivity and dispersivity, which are difficult to measure and spatially variable. The uncertainty in the analysis has not been taken into consideration, therefore just running the model in deterministic mode with the best estimations of the parameters would not provide enough information. The suggested substitute is to employ Monte Carlo analysis or other stochastic approaches to explicitly take into account the uncertainty that exists in any study. It is advisable to promote the regular evaluation of policies and regulations to identify and integrate novel remediation methods and/or advanced modeling expertise. One of the most significant technological drawbacks of current mixing zone models is that they typically fail to take into consideration biological or chemical processes that occur within the mixing zone. Nonetheless, there are a few basic first-order decay methods accessible. Only biological and chemical processes with comparatively short time scales need to be taken into account since time scales inside mixing zones are small. Aquifers are typically thought of as sources of large volume, but the majority of mixing zones have far more restricted spatially. Aquifer time scales are also frequently relatively long, particularly when compared to the majority of mixing zones in ground water-surface water interactions.

9.7. TRANSPORT MODELS

Most hydrologic and hydraulic transport models do not represent stream subsurface interactions. This occurs mostly as a result of the conventional division between modeling groundwater and surface water as well as the particular challenges associated with modeling each of the two hydrodynamic environments. Therefore, in modeling packages like MODFLOW and TOXIWASP, the stream-subsurface contact/interface is regarded as essentially a boundary condition. The primary challenge lies in the fact that these models are unable to compute stream subsurface exchange flux without first defining essential hydrodynamic exchange parameters. Local factors including hydraulic conductivity and sediment size, as well as hydrodynamic conditions on both sides of the interface, affect the exchange flux. More detailed models will need to be created in order to accurately simulate the movement of pollutants across the surface–subsurface contact/interface. But given that not all of the specific hydrodynamic mechanisms that drive exchange are yet fully understood, it looks that developing the necessary degree of modeling sophistication will take some time. Beyond this, the complexity of the contaminant transport problem arises from the need to comprehend the interaction between the transport and transformation processes. A different strategy has been employing groundwater models like MODFLOW to simulate plan-form exchange flows, like those that occur through meanders (Marzadri et al., 2014). While this kind of model works well for simulating larger-scale exchange fluxes, it is constrained by its inability to resolve the fine-scale characteristics and flows that are probably crucial for pollutant outflows. These models are useful because they tackle the issue of contamination transport in great detail, but they are still quite limited because they only look at a small number of exchange events, and scaling to actual environmental contaminant transport presents challenges. There are currently models for the physicochemical exchange of metals, surfactants, and colloids, as well as exchanges caused by bedforms, barriers, and biogenic sediment mounds. A small number of transport mechanisms were the focus of experimental research that produced all of these process-level models. Recently, researchers introduced a new model that models net solute transport in streams. This method offers some insights for using intricate process models on larger sizes. Conjunctive steam-aquifer flow and solute transport simulations were extended to include transitory storage. Many other exchange processes, such as the interaction between physical transport and changes of biogeochemical contaminants, still require investigation. As a result, creating a comprehensive set of process descriptions and incorporating them into a comprehensive model for stream subsurface exchange will take a substantial amount of work. A comprehensive understanding of the hydrodynamics in the mixing zone necessitates a quantitative understanding of the concentrations of ground water as it escapes into surface water. Models of ground water transfer offer a way to advance such knowledge. Generally, the receiving surface water body might be considered a boundary. To model the transport, suitable head and concentration conditions can be applied at this border.

The modeling becomes a little more difficult when tidal oscillations are present in the receiving surface water body. However, because a large number of industries have developed on coastal areas, the problem has high practical significance. The issue is physically complex because, at high tide, surface water from the tidal water body intrudes into the groundwater aquifer to the extent determined by the aquifer characteristics. As a result, there is no discharge 50% of the time. However, pure surface water diffuses into the aquifers and reduces pollutant concentrations, providing that there is no contamination from upstream sources. By diluting the ground water with surface water, this tidal flushing helps to reduce the outflow concentration. A suitable groundwater model can be used to quantitatively estimate the degree of dilution. Groundwater flow velocities in nearby aquifers are sinusoidal due to tides in coastal waters and estuaries. Many researchers have utilized the mathematical relationships to describe the temporal and spatial characteristics of groundwater flow regimes that are close to coastal areas. To solve the transport problem, researchers created a mathematical model using that approach to characterize the fluctuating velocity field. In these investigations, dilution was assessed either directly through the addition of tracers like radioisotopes or fluorescent dye, or indirectly through the observation of salinity fluctuations. Fluorescent dye was introduced to the sewage in many investigations (Carstea et al., 2016).

9.8. MIXING ZONE MODELING

The field of mixing zone modeling is frequently made more difficult by the sudden changes in physical conditions that might occur, for instance, in coastal seas, both in space and time. Under these conditions, recent developments in oceanographic instrumentation promise significantly better modeling. Acoustic doppler current profilers (ADCPs) and thermistor strings are especially useful because they enable continuous assessment of density structure and velocity across the whole water column. Long-term series of observations from these types of devices combined with appropriate mathematical models can provide a great deal of information into the statistical variance of mixing zone parameters. Accompanying the current meter data from ADCPs, coupled near- and far-field models, and density stratification data from thermistor strings allowed for predictions of bacterial dispersion around the outfalls. The optical chemical sensors are now being developed to enable real-time in situ monitoring of hydrocarbons and other naturally occurring species in wastewaters. With the use of these, mixing parameters in intricate surface water ecosystems might be measured without the need for additional tracers. The development of wavelength-definable semiconductor quantum-cascade lasers, for instance, is a recent advancement for mid-infrared applications. This involves infrared diagnostic systems that can monitor absorption at freely selectable wavelengths with exceptionally high sensitivity, enabling the monitoring of specific substances even in the presence of other compounds. To define the basic relationship between the amounts of contaminants in the sediment and the exposure and risk to the environment and human health, a conceptual model is required. The mechanisms that transfer pollutants from sediments to receptor organisms and any ongoing sources that may contribute to sediment recontamination are all identified in a conceptual model. To determine which remedial strategies, have the best chance of successfully addressing the significant pollutant processes, a complete conceptual model is required. The conceptual model serves as the basis for identifying and carrying out site management procedures. A conceptual model can be identified and tested with the assistance of a quantitative model. For instance, the capacity to quantitatively forecast the level of pollution in water and biota based on the identified sources can address the question of whether all sources have been identified. If the observed patterns of contamination cannot be replicated, it may indicate the presence of other sources. The inability to forecast the transfer of contaminants from sediment to water using accepted mechanisms and processes may suggest the presence of other processes at work.

9.9. RISKS

The baseline risks determine the significance of a polluted sediment problem and the necessity of management or remediation within a risk-based decision-making framework (Bridges et al., 2006). While field measurements can detect environmental contamination levels and loads in potentially impacted species, a mathematical model is typically needed to show a cause-and-effect relationship between the two. Without determining and measuring the destiny and transit mechanisms that result in exposure, assimilation, and effect, it is impossible to connect sediment concentrations to hazards. Comparing different contaminated sediment management strategies and evaluating important pathways and processes can be done through quantitative modeling. Active management strategies can be compared against current exposure and risk as well as the anticipated attenuation of contaminants due to these natural processes. Quantitative representations of these processes depend on constitutive relationships and measurements of the parameters within those connections. A number of regulatory programs, such as the total maximum daily load (TMDL) program was applied to contaminated sediments. A TMDL is the highest concentration of a contaminant that a water body can have without degrading its water quality. The different sources of the pollution must be assigned a portion of this TM

DL. A TMDL must take seasonal variation into account and provide a margin of safety to account for uncertainty. The quality of many more bodies of water may be further damaged by the addition of contaminants from polluted sediments or by the impact of sediment processes on the water quality above. Because it establishes the necessary connection between pollution loadings to a waterbody and their impact on meeting water quality standards, predictive modeling is a crucial tool in the establishment of Total Maximum Daily Loads (TMDLs). In conditions like contaminated sediment sites, where it is challenging to determine how much of the pollutant loading can be directly attributable to certain sources, modeling is particularly crucial. Sediment toxicity can also be eliminated by designing TMDLs. While biological testing is a viable method for determining the toxicity of sediment, a numerical/mathematical model is the only tool available for linking point or nonpoint sources of contamination to sediment concentrations. The assimilative capability of the particular receptor at risk, as well as the access and availability of contaminants, must be determinable by mathematical models.

9.10. MANAGEMENT STRATEGIES

Furthermore, to ascertain whether management strategies can completely eradicate sediment toxicity, numerical modeling is necessary. A foundation for assigning permissible pollutant discharges to point and nonpoint sources can also be established using numerical/mathematical models. Protecting resources at risk, such as ecological or human health, commercial or recreational fishing, or a specific endangered species, is the main objective of contaminated sediment management. The ideal course of action would be to choose management and corrective measures that maximize resource protection or facilitate resource recovery. It is necessary to compare the potential future effects of different situations using quantitative models, as measurements can only attempt to depict the current situation. Regarding the optimal way to apply the forward projections in time, there are still important considerations. Models, in theory, give exposure rates or concentrations as a function of location and time. This data offers a foundation for estimating risk and impacts. However, there is a great deal of uncertainty surrounding the estimation of future concentrations and exposure rates as well as the estimation of future hazards and impacts. Using pollutant flows as a substitution for exposure and, eventually, risk is an approach to assessing and contrasting treatment strategies. If a technology releases fewer pollutants into the air and water and leaves less residual pollution than an alternative technology, then it can typically be believed to pose less exposure and risk. By offering a screening technique, the assessment of pollutants for each treatment strategy can be especially helpful in the comparison. The elements of a comprehensive management plan that primarily regulate the overall exposure or risk can also be found through a comparative study of pollutant flows. As a result, these elements should be given more attention and resources for a thorough assessment. Screening sediment management options can guarantee that all necessary elements of a specific option are included for further assessment in this way. In general, complex modeling of the fate and transport processes of contaminated material is necessary even for a comparative study of mass flow. Even though pollutant flow can be helpful, exposure and risk to ecological and human health ultimately determine if remediation is necessary and whether any management or remedial strategy is successful. Particularly when it comes to balancing acute, short-term risks with long-term ones, pollutant flow is not particularly useful. For example, the removal of contaminated silt often results in higher short-term risks in return for possibly lower long-term risks. It is crucial to remember that methods for in situ sediment management are always vulnerable to possible setbacks due to unanticipated conditions. A foundation for determining the extent of future exposures and risk that the different management approaches may present can be found in modeling.

9.11. SEDIMENTS

Due to their strong associations with solids, the majority of important pollutants and other contaminants of concern (COCs) tend to collect in sediments (Olutona et al., 2016). Since they are efficiently released into the surrounding water, pollutants such as soluble metals and polar organic molecules that do not strongly bind with particles are rarely considered sediment contaminants. Sediment contamination is frequently caused by surface runoff and industrial and municipal wastewaters, as only pollutants that have a significant tendency to partition to the solid phase have remained in their actual locations. Pollutants that are soluble and volatile are typically carried away by flowing water or released into the atmosphere through evaporative processes. Similar to this, when more soluble and volatile contaminants first pollute sediments, they can eventually move into more mobile phases due to their mobility, which removes the contaminants from the sediments. The majority of pollutants in sediment are those that have a strong association/bonding with the solid phase, however there are certain exceptions to this when contaminants are still being added to the sediments from ground waters or active sources. However, the high correlation between contaminants and the solid phase in a sediment bed indicates that sediment mobility frequently determines the fate of contaminants. Antimony, arsenic, beryllium, cadmium, copper, lead, mercury, nickel, silver, thallium, and zinc are among the hazardous elements. The fact that these contaminants are poisonous in solutions, biomagnified, and nonbiodegradable makes them significant. Many of these substances have complicated chemistry in sediments. Without chemical changes in the sediment, a portion is usually chemically fixed and largely inaccessible to fish and higher species. A section that is frequently ion exchangeable can be made available by adding a more firmly held contaminant.

The leachable percentage of metals is normally less than 10% and occasionally much less because the partition coefficient between the leachable fraction and the water is typically between 3 and 10. The pH and oxidation-reduction conditions of the water and sediment, in particular, determine the equilibrium state for metals and other species. For metals, the ratio of equilibrium water concentration to sediment loading is frequently very considerable, yet only a tiny portion of the metals are usually available. Because of this unpredictability, a measurement of the sediment-water partition coefficient specific to a given location is favored above any forecasting method. Many metals' availability seems to be influenced by the existence of acid volatile sulfides (AVS) in the sediments where they are found. The method used to assess sulfide presence is referred to as AVS. Most of the major metals of concern are linked to the sediment as sulfides if the ratio of AVS is higher than the amount of simultaneously extractable metal (SEM). The metals seem to be mainly inaccessible to receptors in this form, ranging from bacteria to humans. Significant amounts of AVS are present in many marine sediments, thus it's possible that most of the metals aren't accessible to organisms for exposure. Low AVS/SEM ratios may not always imply accessible metals, even though high ratios suggest that some metals are bound in forms that are not available (Nizoli & Luiz-Silva, 2012). Traditionally, the most significant sources of these elements were runoff from urban and rural areas and the metals processing industry. Tetraethyl lead, which is added to gasoline to prevent premature ignition, also contributed significantly to the environmental distribution of lead. Certain metal sources, such as leaching from abandoned mining sites and urban runoff, are poorly regulated even if ongoing sources have been largely contained. In contrast to pollutants that require oxygen to be present, these pollutants are difficult for natural mechanisms to eliminate. During chemical fate reactions, a range of organic and inorganic molecules in sediments use up oxygen. Sediment oxygen demand, a metric that bears significance akin to oxygen-demanding parameters in the surrounding water, is utilized to quantify the combined impact of their existence. The purpose of sediment oxygen demand is to lower the amount of oxygen that is accessible in the sediment and promote anaerobic conditions there. This may affect the chemical state of metals, so limiting their mobility, as well as the rate at which fate processes (biological contaminant degradation) occur. The oxygen demand of the sediment in slow-moving water can also affect the oxygen content of the water above it. No specific concentration of components that require oxygen is deemed problematic. Instead, the dynamics of the sediment and water column above them determine how these toxins affect the environment.

9.12. GREASE AND OILS

Grease (Figure 9.1) and other long-chain nonpolar organic chemicals have a strong affinity for solids and sediments. Total petroleum hydrocarbon (TPH) concentration and oil and grease content are used to quantify their presence in sediments (Quiñonez-Plaza et al., 2017). These chemicals are typically sourced from TPH production or processing plants, as well as facilities that utilize or process substantial quantities of these compounds. Significant accumulation in sediments can also be caused over time by wastewater treatment effluents from industry and municipalities. Grease and oil or TPH levels in sediments are frequently good markers of past pollution because many of these sources are more strictly regulated than in the past.

Complex combinations of exceptionally stable organochlorines, PCBs are widely employed in industry, particularly as transformer and electrical capacitor lubricants. PCBs' durability also contributes to their environmental persistence. PCBs are easily absorbed by aquatic animals, soluble in body lipids, and biomagnified in the food chain, just as organochlorine herbicides. Important health concerns were generated by specific isomers and trace contamination with other chlorinated species, despite the comparatively low toxicity of many individual PCBs.

Figure 9.1. *Grease.*

Source: https://commons.wikimedia.org/wiki/File:Grease_trap_for_graywater_(5293658840).jpg, Creative Commons License.

Consequently, the United States banned the manufacture of PCBs in 1979. It is important to note that PCBs are a complicated mixture of chemicals and are typically only identified by the amount of chlorine they contain overall. Aroclors are specific mixes of PCBs. PCBs are mostly stored in industrialized regions in the northeastern United States and the Great Lakes. Many of the Great Lakes have fish warnings in place due to health risks associated with consuming fish contaminated with PCBs. PAHs are found in fossil fuels and are employed as chemical intermediates. Examples of these include naphthalene, pyrene, and chrysene. PAHs with two aromatic rings, like naphthalene, are often the most soluble, mobile, and volatile. The number of rings tends to reduce solubility, volatility, and degradability under natural conditions. Because PAHs are less volatile than monocyclic aromatics, they are more likely to be found in coal tar and heavier oils than in lighter oils. Certain PAHs are thought to cause cancer in people as well as animals, where they have been shown to cause cancer. PAHs have a bioaccumulation potential between PCBs and monocyclic aromatic/halogenated aliphatics, with an intermediate persistence. PAHs are used in the burning of industrial fuels and oils, which has led to their accumulation at closed industrial sites where pollution levels can be very severe.

Because stream velocity is so important to the transport of sediment, a hydrodynamic model needs to accurately represent fluctuations in stream velocity in both space and time. Since simulated conditions are likely to extend, a flexible sediment transport model cannot be based solely on measured velocity fields. The streambed geometry and bathymetry, flow rates, downstream water surface elevation, and water and atmospheric parameters are among the data needed for such a hydrodynamic model. The model can be validated or calibrated using measurements of the current velocity and height of the water, as well as properties. Calibration and validation data are obtained from measured suspended solid concentrations and changes in sediment bed elevation. In order to maintain consistency, bathymetry and bed variability should be resolved using the same grid with sufficient resolution in both hydrodynamic and sediment transport models. Material balance-based models should be used for both sediment transport and hydrodynamics. The system determines the necessary dimensions as well as the level of complexity in the wind-forcing model. To facilitate extrapolation outside of the calibration validation database, the sediment transport model needs to be fundamental and mechanistic. The sediment movement model should track noncohesive sandy components and flocculating clay and silt independently. The mechanics of resuspension and settling are not captured by models that solely replicate one size category of particles, making them unsuitable for accurately correlating sediment and pollutants. The origins and sinks of specific sediment deposits must be uniquely identified in order to predict the dynamics of contaminants. Dynamics of contaminants can be predicted even when total suspended sediment loads appear to be reproduced rather accurately. Shear stress and, in the case of cohesive sediments, sorption effects should be taken into consideration when describing the dynamics of sediment and contaminants (Ongley et al., 1992).

9.13. SEDIMENT TRANSPORT

It is important to stress that the destiny/fate of contaminants has no bearing on sediment transport. It is not appropriate to calibrate the sediment transport model using observations of pollutant fate and movement. Rather, an extra limitation on the pollutant transmission is given by a sediment transport model. The conventional paradigm postulates that sediment dynamics substantially determine the destiny and behavior of hydrophobic and other pollutants that are closely connected with the solid phase. The local solid phase density and the effective sediment-water partition coefficient determine how much of the pollutants are linked to the solid phase. Even after dredging, suspended sediment loads in rivers seldom surpass 1 g/l. In water with reduced quantities of suspended sediment or sediment-water partition coefficients, the pollutant is mostly present in the dissolved state. Under such conditions, other processes, such as vaporization and in-stream dispersion, dominate contaminant transport, and the fate of the contaminants is no longer linked to the fate of the sediment. The vital connection between sediments, water, and biota is provided by the sediment bed surface layer. Pollutants from the surface layer are accumulated by invertebrates that reside in or on the sediment bed. Chemistry and sediment dynamics work together to regulate the amount of contaminants in the top layer. The pace at which contaminated surface sediments are buried, or removed from the active surface layer, is determined by the rates of deposition and resuspension. Furthermore, the degree to which buried contaminants are kept accessible to the surface layer is determined by the depth and extent of surface layer mixing. The particle matter in the bed is more closely linked to chemicals that are comparatively more hydrophobic. As a result, there are fewer concentrations dissolved in the pore water and less concentrations in the overlying water column accessible for diffusion and advection. It is possible to separate the dynamics of contaminants from hydrodynamics, sediment deposition, and erosion under certain conditions. When bulk sediment movement is irrelevant, a number of processes, such as diffusion, advection, and bioturbation, may be responsible for the exchange of contaminants. In these conditions, a fate and transport model inside the water column is bound by pollutant exchange at the sediment-water interface. The sediment boundary must be identified as a source or sink of contaminants in order to calculate the concentrations of contaminants in the water column as well as the fate and movement of contaminants carried by water.

Either the movement within the sediment bed or the movement within the benthic boundary layer in the overlying water can serve as the regulating barrier to contaminant mass transfer to or from stable sediments. Simple lumped exchange coefficients are usually utilized to characterize contaminant release from the sediment since contaminated sediment models have focused on accurately representing the sediment transport. The most fundamental and common chemical transport mechanism in a sediment bed is molecular diffusion. Contaminant molecules migrate from areas of high porewater concentration to areas of low concentration as a result of random molecular motion in pore water. Additionally, Darcy velocity operates on pollutants present in porewater. There are several sizes at which hydraulic exchange activities take place at the sediment-water interface. Local pressure changes can accelerate in-bed migration and pollutant transfer on the scale of the uneven bedforms that usually characterize the bottom surface. Similarly, local advective flows within the underlying sediment can be produced by cross-stream pressure fluctuations at stream meanders. The net groundwater migration into or out of the stream may be caused by regional groundwater gradients. Significant differences in the magnitude or even direction of this transport can also result from local variations in bed characteristics. It is challenging to estimate ground water flow velocities because they must account for the seasonality and spatial variability of groundwater flow, particularly stream bed seepage velocities. Seepage meters, which are essentially containers that cover a section of the sediment bed and collect water that seeps through the sediment, are frequently used. However, because the majority of any seepage that is recorded may occur via sandy areas of the bed that contain little to no contamination, net seepage data may not provide a complete picture of contaminant transportation. Researchers reported that the net effect of seepage on pollutant movement has been estimated using chemical tracers in the overlying water. The fact that advection may also be influenced by localized changes in pressure on the irregular surface of the sediments adds complexity to spatial variability (Bodmer et al., 2020).

9.14. GROUNDWATER

Measuring the ground water flow in the surrounding aquifer, as opposed to velocity, might be helpful since it shows the average in- or out-flow from the waterbody. Piezometers positioned at various elevations beneath the bed can measure the general direction of the groundwater flow. Inflow happens when the head of water beneath the surface is higher than the head in the stream' outflow occurs when the opposite is true. If the permeability can be determined, this data is also utilized to estimate the flow rate in addition to direction. Tracers offer an additional method of identifying slow vertical transmission by groundwater flow. Preferred water flow and pollutant movement can be facilitated by pathways found in animal burrows and root systems. More importantly, benthic animals like worms frequently stir the near-surface sediment continuously. Animals living in the higher layers process sediment through burrowing, ingestion and excretion, tube construction, and biodeposition. When combined, these phenomena are referred to as bioturbation. Particles of sediment and pore water move both vertically and horizontally as a result of bioturbation. The bioturbation process also transfers pollutants that are on the particles or in the pore spaces. This is crucial when transferring hydrophobic contaminants that have been significantly slowed down by pore-water activities. Advection-based pollutant transfer at the sediment-water interface can be significantly enhanced by worm tubes and other macroscopic animal burrows. Furthermore, consuming sediment deposits directly may cause the sediment and related pollutants to rise quickly to the surface. The majority of contaminated sediment models do not completely account for bioturbation and frequently include it with other processes in the sediment-side mass exchange coefficient or the effective biodiffusion coefficient. Since effective biodiffusion or exchange coefficients cannot distinguish between the several particle movement modes that an organism exhibits, they are imprecise approximations. Furthermore, pore-water pumping and the impact of bioturbation on other diffusive and advective processes are not included because they are often studied with strongly sorbing pollutants.

9.15. DIFFUSIVE TRANSPORT

Diffusive transport is around 30 times smaller than bioturbation-induced transport. The increase caused by bioturbation may be negligible for certain species whose leachable fraction can only partially partition into the solid phase. For some species, pore-water pumping or other soluble transport mechanisms that are not included in the majority of biodiffusion coefficient measurements may be crucial. The gas usually originates from biological processes, such as the anaerobic metabolism of microbes producing methane or hydrogen sulfide. At the gas-solid interface, hydrophobic organic molecules have a tendency to sorb more selectively. The bulk movement of the gas phase may theoretically change the physical characteristics of the sediment bed, and even a relatively tiny amount of pollution flowing in this way could lead to surface recontamination. In the vicinity of individual animal burrows, bioturbation and other biological activities dominate movement. Unless the sediment composition or degree of contamination prohibits significant colonization by benthic organisms or unless groundwater seepage is so high that advection prevails, bioturbation is the main migration mechanism of sorbing contaminants in stable sediments. The impact of bioturbation is typically included in an overall mass exchange coefficient along with other sediment-side transport mechanisms in the models of contaminant transfer from sediments. In cases where hydrophobic chemicals are closely linked to particulate matter, other pathways take precedence. The variety of species, their feeding habits, and their migration patterns confound the investigation of pollutant transport and exposure via the food web. The main organism of interest is frequently positioned high in the food chain, with several tiers/stages dividing it from the source of contamination. The food webs change with space and time, particularly with regard to season. One way to define a biota and sediment accumulation factor (BSAF) is the proportion of pollutant concentrations in the local sediments to the organisms (Kwok et al., 2013). When it comes to hydrophobic substances, an organism's lipid composition frequently determines its propensity to accumulate pollutants.

9.16. ORGANIC CARBON

Similar to this, the organic carbon content of the sediment greatly influences the bioavailability of the pollutant in the sediment. Therefore, the ratio of the lipid accumulation in an organism to the organic carbon quantity in the sediment is commonly defined as the BSAF for hydrophobic substances. Explicit modeling of contamination transport through the food web is required to enhance the capacity to forecast exposure and accumulation at higher trophic levels. There are bioaccumulation models that incorporate complex movement patterns within a dynamic framework to simulate the transmission of pollutants across intricate, dynamic food webs over the course of an organism's life cycle. The USEPA used the model framework to construct water quality criteria, and it has also been implemented at many sites with multiple contaminants. It was discovered that many models compute identical pollutant concentrations when they were recently compared. Some researchers used the age-dependent, time-varying bioaccumulation model to forecast PCB transport via a food web consisting of striped bass, white perch, phytoplankton, and zooplankton in the lower Hudson River. At each of these trophic levels, time-dependent diffusive uptake, ingestion, egestion and excretion, growth dilution, and metabolic activities were modeled. The field of bioaccumulation modeling has advanced to a point that it is now feasible to accurately calculate the trophic transmission of numerous compounds, such as hydrophobic organics and metals, in aquatic food webs. The data that characterize biological processes, specifically food web structure and the ultimate source of contamination and movement/migration patterns, are typically the limits of such models. Additionally, food web models can be connected to chemical fate and transport models to provide a unified mass balance framework for evaluating possible remediation techniques.

The analysis is not finished by defining exposure and accumulation in receptor organisms. Translating exposure and uptake into impacts is ultimately necessary for a risk-based approach to controlling contaminated sediments. Relative exposure estimates might be helpful when contrasting different management options. Determining the true risks of doing nothing, however, is what determines if any management action is necessary. The degree to which a management activity achieves its objectives for risk reduction also determines its effectiveness. Toxicological information is typically restricted to estimations of impact threshold concentrations of specific reactions in specific species when calculating risk to ecological receptors. Usually, a conservative approach is taken when applying such threshold concentrations. In addition, threshold values derived from laboratory toxicity studies typically do not directly address the problem of population-level consequences, even while they give an indication of the potential for effects on people. Toxic equivalency factors are important for complex combinations of endocrine disruptors. However, the usefulness of toxic equivalence factors is limited to similar species displaying similar effects, that is, to those whose mechanism of action and reaction are identical to the method used to determine the risk equivalency. Understanding the consequent uncertainty in dose-response or effects thresholds between species and among individuals within a species is crucial. Using a population or ecosystem model in conjunction with a comprehensive dose-response relationship, rather than just one threshold concentration, is a more advanced method of assessing ecological risk. The field is expanding as a result of the relationship between environmental toxicology and ecology. Methods include individual-based models, life-history models, scalar abundance models, and metapopulation models with the aid of geographic information systems. A crucial aspect of modeling ecological effects involves simulating the environment or terrain in which the ecosystem exists. The ecosystem and its response to contaminants are evidently altered by dynamic changes in the landscape, such as seasonal variations. A single threshold dose, or the level below which risks are acceptable, is typically used in human health risk assessment. Critical dosages for consequences other than cancer are determined separately based on the chemical. The focus of human health risk assessment is on individual protection, which sets it apart from ecological risk assessment. With the exception of endangered species, the protection of populations is the primary goal of ecological risk assessment. The threshold dosages as well as food consumption rates and patterns are important sources of uncertainty. Furthermore, as humans typically eat apex predators, the uncertainties surrounding the bioaccumulation calculation add to the overall uncertainty surrounding the risk assessment for human health.

9.17. MODEL

Every model has uncertainty and error, but these don't present problems if they're small or related to unimportant problems in relation to the questions the model is meant to answer. Models are created with certain goals in mind, and their appropriateness and suitability are determined only by how well they accomplish those goals. Evaluating uncertainty is a crucial part of any model. Rather, distinct realizations of a model might arise from varying assumptions and model formulations, without significantly affecting the model capacity to replicate data. The conclusions that are derived from the model's future projections may vary significantly. Either a conceptualization that is not correct or a conceptualization that is not well translated into a numerical model might be the cause of equation error. The inability of models to identify source and sink of contaminants poses an obstacle to the description of present conditions and the exploration of sediment management strategies. When such a model is calibrated to data, it may lead to erroneous conclusions on the presence or significance of other processes in the model and the most effective way to handle the risks those processes present. Similar issues may arise from improper formulation of particular processes. Errors can be found by comparing data patterns with model projections. The adoption of methods that show excessive numerical diffusion or instability, as well as spatial and time-grid discretization errors, are examples of numerical implementation problems. In contrast to process description flaws/errors, these issues may be assessed or minimized using conventional methodologies. Inaccurate parameters or parameterizations might result from calibrating to data that is erroneous or insufficient for system characterization. The constitutive relationships and parameters that utilize them are typically applied to field or laboratory data within particular settings. Error can occur when these parameters are used outside of specified ranges. It is impossible for parameters to be independent, and calibration measures reflect this covariance, making parameter calibration more challenging and less accurate. Alignment data gaps and uncertainties further restrict parameter accuracy.

Models are used to project future conditions in order to assess the efficacy of natural recovery or alternative polluted sediment management strategies. Future conditions must be assumed, but baseline conditions that are required to support a model can be measured. Although they may not represent model errors, these assumption errors do reduce the accuracy of predictions. Specifically, assumptions about future conditions have the potential to erroneously influence modeling results and unfairly accept or reject particular solutions for sediment management. A model acceptability is determined by the questions posed of it and the accuracy needed in the responses. The model's major objective is frequently to offer a foundation for selecting an effective environmental management approach to preserve or enhance the quality of the water or sediment. The model's capacity to distinguish between different management strategies may be hampered by uncertainty. The controllable uncertainty is usually restricted to the model parameters, provided that the model is a potentially realistic description of the system and that the future conditions that need to be predicted are sufficiently well defined. Certain parameter values can be determined with extreme precision. The number of parameters that are likely to be specified in this way is increased when fundamental and theoretically valid parameterizations or constitutive relationships are used. Field data must be used to calibrate the remaining values. Therefore, the characteristics to which the model is least sensitive throughout the calibration process are also the most uncertain. This uncertainty may result in model limits if the model is later required to predict conditions outside of the calibration range. Typically, the process of evaluating model uncertainty involves generating model outcomes across a spectrum of parameter values. Bayesian analysis might be used to assess the distribution of the model outputs if the parameter value distributions and the covariance between them are known. Unfortunately, with sophisticated models of polluted sediment, such an approach is practically unfeasible. In most cases, the parameter distributions and their covariance are unknown. Monte Carlo simulations of model uncertainty frequently employ randomly chosen parameter values. However, such a method can result in false findings if the parameter distributions and/or covariance between parameters are unknown. These boundary estimates can be applied immediately to analyze the model sensitivity to parameter changes or to investigate the boundaries of the model predictions. In order to make sure that unknown parameter covariance does not invalidate the results, bounding estimates must be employed carefully. Subjectivity is inevitably introduced when determining parameter boundaries using engineering judgment. Bounding estimates should therefore be chosen with care, sufficiently backed by documentation, and subjected to critical evaluation. Bounding estimating is frequently the only practical way to investigate model uncertainty due to parameter uncertainty.

When two or more conceptualizations of a system or management approach have been identified, an alternate strategy for investigating model sensitivity is to try to support the alternative conceptualization or approach within a fair range of parameter values. The degree of confidence in the inferences made from the model determines its acceptance/selection. Although there isn't always a suitable quantitative measure of uncertainty, there are some that can be helpful. The RMSE or root mean squared error between model predictions and observations can be both helpful and misleading. Neither bias in the model predictions nor agreement with data trends are indicated by the root mean squared error. Additional factors like bias and agreement with temporal and spatial trends should be assessed. All models should give a conclusion of the material balance since fundamentally sound models are based on a material balance inside the system. Inaccuracies in the model's conceptual or numerical formulation may be the cause of the failure to determine the material balance. The degree of confidence with which the model results should be assessed can also be determined by comparing the sensitivity of the model predictions with parameter values and with estimates of the uncertainty in those parameters. Compared to sandy sediments, cohesive sediments require a higher friction velocity for a given particle size in order to cause particle motion. Cohesiveness is a complex function of salinity, mineral concentration, particle size, bulk density, and organic content (Grabowski et al., 2011). Sediment transport models currently face the issue of incorporating these variances and strengthening the theoretical foundation for predicting cohesive sediment motion. The excess reliance on stream velocity highlights how important it is to model stream hydraulics at high precision when attempting to forecast the movement of silt. Without bed data, erosion prediction is still not achievable, even with the stream hydraulics. Because of the impacts of bed shear stress, cohesiveness of the depositing sediment, and flocculation processes, deposition varies significantly over time and space. As salinity, sediment concentration, and fluid shear rise, floc size and settling speed decrease. Compared to flocs created under various circumstances, flocs produced at lower fluid shears and sediment concentrations have a tendency to settle more slowly. When fluid shear is minimal, flocs generated under differential settling conditions tend to approach steady-state size distributions more slowly and settle faster than flocs affected by significant fluid shear. Although difficult to forecast, these differences in settling and depositional dynamics from flocculation are essential to accurately characterize sediment erosion and deposition. The density and erodibility of sediment beds are also influenced by processes like gas production.

9.18. DATA COLLECTION

Measurements involving sediments from the Detroit River taken over a two- to three-month period show that the attributes of sediment beds are also time-dependent. The bulk density went from 1.38 to 1.48 g/cm^3, changing the erodibility by two orders of magnitude. Changes in the characteristics of the depositing particles over time may also have an impact on the resultant bed armoring. Experimental methods such as Sedflume can be used to acquire erodibility data with shear stress and sediment density, even though a priori prediction is not feasible. To keep the bottom surface of the Sedflume device level while the sediment erodes, the sample to be analyzed is lifted continually. Some models of sediment flow can be derived from the fundamental constitutive relationships derived from these measurements. To enable high-resolution sediment transport models that take into consideration the physical property variabilities of the sediment beds with respect to time and place, extensive data are required. Better bed representations are required since sediment erosion depends on bed characteristics and consolidation. The bulk density is either constant overall or constant within certain layers in the Recovery and WASP5 models. There is a variance in the bulk density in SED2D, HSTCM, and CE-QUAL-ICTOX. Nonetheless, because the bed characteristics are defined separately, the pore-water flow may not always align with modifications in the bed properties. The USEPA created the Environmental Fluid Dynamics Code, which is intended to provide a unified framework for simulating sediment transport, hydrodynamics, and contaminants. Similar to SED2D, the void ratio is determined by a specific depth-dependent profile' however, the prediction of bed parameters with respect to time and depth is based on strain consolidation. This allows for the estimation of the bed dynamic response to intermittent flow episodes. When assessing in situ contaminated sediment management alternatives, such as natural recovery and in situ capping, the definition of bed dynamics and bed response to flow episodes is crucial. Significant effects are also influenced by high-frequency variations in flow (turbulence) on erosion and deposition behavior.

9.19. TURBULENT FLOW

The turbulent flow structure is influenced by vessel movements, and this has an impact on the sediment resuspension. The influence of flow on sediment resuspension is not entirely captured by the average bed shear stress. There is a phase lag between bed erosion and velocity fluctuations that is depending on particle size. Erosion rate estimates from models can only be made when they are grounded in observations unique to a given site and associated with bed and flow characteristics through the use of physically grounded, theoretically sound models. Any confidence that the model can be extended beyond the range of collected measurements or utilized to forecast future behavior comes only from a strong theoretical foundation. To do this, more accurate models of cohesive sediment properties, their temporal and spatial fluctuations, and their interactions with the hydraulics of the overlying water are required. Many biological processes are grouped together under general exchange coefficients or under simple accumulation metrics like the biota-sediment accumulation factor. Enhancing the capacity to simulate the fate of contaminants in sediments necessitates concentrating more on specific organisms and processes. It is necessary to have a better understanding of how different groups of organisms interact with sediments and to create models that more accurately depict these interactions. Dynamic food webs of important target species must be incorporated into models in order to improve accumulation estimates above BSAF computations (Carbonell et al., 2000). The ecological risk assessments must be expanded to include the entire ecosystem as part of the modeling attempt. Creating stronger connections between contamination modeling and ecological dynamics and consequences is necessary since risk minimization is the core idea underlying the management of polluted sediments. The availability and release of metals are generally far more complex than those of hydrophobic organic processes. A large portion of our existing knowledge is restricted to the thermodynamics of a certain class of pollutants found in sediments. Research tools should be used on field probes to solve the issue and improve the definition of field conditions. Moving beyond equilibrium predictions, attention must also be paid to the dynamics of metal speciation, reaction, and transport. At this point, it is particularly challenging to describe multisolute, multisorbent effects. Transitions between anaerobic and aerobic conditions can have a substantial impact on the release of metals, and biological processes are frequently the driving force behind metal speciation. Therefore, a deeper understanding of the connection with biological processes is necessary. This is especially true for mercury, which participates in important biologically mediated reactions in the environment. AVS can affect the availability of arsenic and mercury in addition to various other metals, especially divalent metal ions, for which it has been demonstrated to have a significant impact.

9.20. HOC

Unlike many metal species, hydrophobic organic compounds (HOC) are not susceptible to the intricate processes of oxidation, reduction, and other reactions. A significant portion of HOC pollutants may not desorb from sediments, or may only desorb to a limited extent or over a long period of time. Research utilizing various pollutants and sediment systems has produced findings that seem to be different. Predicting the amount and rate of desorption from contaminated sediments is necessary, and developing those predictions will require additional research.

It's important to ascertain if rate-limited or bimodal HOC desorption occurs and how it affects contaminant availability and uptake. Given the heterogeneity of hydrogeologic and pollutant settings of hazardous waste sites, designing and sustaining efficient remediation systems that satisfy all technical, regulatory, and societal limitations is an incredibly difficult task. For all stages of site remediation, the USEPA encourages the use of the most innovative, practical, and efficient tools available, ranging from cutting-edge, field-based technology for site characterization to enhanced data management and visualization tools to cutting-edge in situ and ex situ remediation methods. Mathematical optimization for the design and redesign of remediation and monitoring systems is one such potential new technology. However, due to a lack of cost and performance statistics as well as concerns about regulatory approval, the regulatory community has been hesitant to adopt these ideas readily, as has been the case with many revolutionary technologies. A demonstration project utilizing hydraulic optimization approaches for pump-and-treat systems was done by the USEPA in 1999.

The objectives included identifying three locations with operational pump-and-treat systems, evaluating each location for possibilities for optimization, and implementing a hydraulic optimization code at each location. Pumping options that could save millions of dollars in comparison to present pumping rate expenses were found at two of the sites. With optimization, no discernible improvement over the design was found at the third location. The conclusions were that larger groundwater plumes and many extraction and injection wells are examples of complex sites that may benefit more from hydraulic optimization, and that hydraulic optimization itself has the potential to enhance the performance of pump-and-treat systems. It is important to remember that while there are other mathematical optimization techniques accessible, this study only assessed one method of hydraulic optimization.

Many problems, such as a lack of understanding, a shortage of skilled optimization modelers in the consulting engineering community, and expense, can be the reason for the underuse of optimization techniques. The remedial community's ignorance of optimization approaches is undoubtedly the main cause of its inadequate adoption. Even while optimization algorithms are widely used in many different industries, these methods are not yet considered standard practices in the remediation field. Moreover, there aren't many experienced users or practical applications. Because of this, the optimization algorithm benefits are not well understood by industry and government sectors, and they lack professionals with the necessary training for these applications. From a regulatory standpoint, authorities have not been extensively exposed to the applications of mathematical optimization algorithms as few sites have requested to utilize them.

Cost is another issue linked to the underutilization of mathematical optimization. Many researchers have created basic flow models using limited site characterization data. These models are frequently insufficient for use with an optimization method, but they are typically employed as a tool in the site decision-making process. There may be additional costs associated with updating or fully redoing the basic model(s) in order to ensure a valuable optimization analysis. There is still a critical need to enhance remediation systems using mathematical optimization or other methods, despite the fact that there are a number of reasons why optimization is not used in the remediation field. It is projected that many of these systems would run for years or even decades at significant expense to the government and business. In addition, a lot of these systems were created with limited site data and a limited understanding of pump and treat systems' capabilities. In these situations, applying mathematical optimization approaches can greatly benefit all stakeholders. It is also advisable to take into account additional continuous improvement strategies like data management, labor and monitoring procedures, aboveground treatment components, and routine system performance evaluations. Pump-and-treat systems and other restorative systems must be carefully planned, maintained, and observed' the remediation community should take optimization techniques into consideration to achieve this goal.

The main goal of industry is to lower its long-term cost liability. Because of this, industry is typically ready to pay for an optimization analysis up front if it is anticipated that doing so will lower the overall life-cycle cost. Prior to conducting a thorough optimization evaluation that takes into consideration the anticipated costs of the optimization analysis, the anticipated costs of system modifications, and the anticipated savings, a site-specific cost-benefit analysis is necessary. The industry typically uses a discount/subsidized rate to convert future expenses to their present value when performing cost evaluations in terms of net present value (NPV). As a result, industry optimization evaluations must be conducted using net present value (NPV). Regulatory bodies must give the industry their approval before industry can adopt or alter corrective measures. In this sense, strategies derived from groundwater modeling alone are no different from those derived from mathematical optimization techniques linked to groundwater simulation models. This is because the mathematical optimization algorithms just run a sequence of simulations with the ground water model in an efficient manner.

Thus, the reliability of the predictions made by the groundwater model should be the main focus of regulatory issues. The industry is well aware that novel strategies for remediating groundwater are constantly developing. In addition to weighing the possible advantages of the current remediation strategy, the decision of whether to use simulation optimization techniques must take into account whether resources would be better used investigating alternative remedial approaches that could supplement or replace the current remediation strategy. Long-term groundwater monitoring expenses may actually represent the main lifecycle cost element at many locations. The biggest potential for future cost savings at these locations may lie in optimizing groundwater monitoring. Optimization management models have been created recently to create groundwater remediation plans. To find the best remedial design, these models integrate mass transport and ground water flow simulators with mathematical optimization approaches.

9.21. MONITORING

Long-term ground water monitoring design optimization is another significant field of research and development that has seen significant growth in recent years (Reed et al., 2013). The longevity of monitoring programs, the requirement to validate treatments, and the possibility of remedy modifications if either the solution or the monitoring plan do not function as intended make long-term groundwater monitoring a significant concern.

It is advised that future study concentrate on monitoring networks that function effectively over extended periods of time, as they have not been well studied before. Early in the 1970s, ground water simulation models were developed, giving planners quantitative methods for examining different management approaches. In order to determine the optimal management options while taking management goals and constraints into account, simulation models and optimization models have been merged in recent years. Simulation optimization models will therefore inherently be somewhat inaccurately specified. The next step is to create remediation design optimization models that guarantee the satisfaction of remediation performance criteria, even in cases where there is a risk assessment. Engineering design has historically depended on the application of predefined design codes that define deterministic safety factors to take uncertainty in the design process into consideration. But when it comes to ground water treatment, every issue is different and there is either no precedence for the problem or very little on which standardized safety criteria may be developed.

BIBLIOGRAPHY

1. Abdollahi, S., Raoufi, Z., Faghiri, I., Savari, A., Nikpour, Y., & Mansouri, A. (2013). Contamination levels and spatial distributions of heavy metals and PAHs in surface sediment of Imam Khomeini Port, Persian Gulf, Iran. *Marine Pollution Bulletin, 71*(1), 336–345. https://doi.org/10.1016/j.marpolbul.2013.01.025.
2. Alshawabkeh, A. N., & Maillacheruvu, K. (2002). Electrochemical and Biogeochemical Interactions under dc Electric Fields. In J. A. Smith & S. E. Burns (Eds.), *Physicochemical Groundwater Remediation* (pp. 73–90). Springer US. https://doi.org/10.1007/0-306-46928-6_4.
3. Alshawabkeh, A. N., Sheahan, T. C., & Wu, X. (2004). Coupling of electrochemical and mechanical processes in soils under DC fields. *Mechanics of Materials, 36*(5), 453–465. https://doi.org/10.1016/S0167-6636(03)00071-1.
4. Awal, M. R. (2009). Environmentally conscious petroleum engineering. In *Environmentally Conscious Fossil Energy Production* (pp. 1–86). John Wiley & Sons, Ltd. https://doi.org/10.1002/9780470432747.ch1.
5. Bakatula, E. N., Richard, D., Neculita, C. M., & Zagury, G. J. (2018). Determination of point of zero charge of natural organic materials. *Environmental Science and Pollution Research, 25*(8), 7823–7833. https://doi.org/10.1007/s11356-017-1115-7.
6. Baldocchi, D., Valentini, R., Running, S., Oechel, W., & Dahlman, R. (1996). Strategies for measuring and modelling carbon dioxide and water vapor fluxes over terrestrial ecosystems. *Global Change Biology, 2*(3), 159–168. https://doi.org/10.1111/j.1365-2486.1996.tb00069.x.
7. Basta, F. F., Maurice, A. E., Bakhit, B. R., Azer, M. K., & El-Sobky, A. F. (2017). Intrusive rocks of the wadi Hamad area, north eastern desert,

Egypt: change of magma composition with maturity of Neoproterozoic continental island arc and the role of collisional plutonism in the differentiation of arc crust. *Lithos, 288–289*, 248–263. https://doi.org/10.1016/j.lithos.2017.07.011.

8. Bayer, P., & Finkel, M. (2006). Life cycle assessment of active and passive groundwater remediation technologies. *Journal of Contaminant Hydrology, 83*(3), 171–199. https://doi.org/10.1016/j.jconhyd.2005.11.005.
9. Bergaya, F., & Lagaly, G. (2013). Chapter 1–General introduction: Clays, clay minerals, and clay science. In F. Bergaya & G. Lagaly (Eds.), *Developments in Clay Science* (Vol. 5, pp. 1–19). Elsevier. https://doi.org/10.1016/B978-0-08-098258-8.00001-8.
10. Bertoni, C., Lofi, J., Micallef, A., & Moe, H. (2020). Seismic reflection methods in offshore groundwater research. *Geosciences, 10*(8), Article 299. https://doi.org/10.3390/geosciences10080299.
11. Bijay-Singh, & Craswell, E. (2021). Fertilizers and nitrate pollution of surface and ground water: An increasingly pervasive global problem. *SN Applied Sciences, 3*(4), 518. https://doi.org/10.1007/s42452-021-04521-8.
12. Biswas, A. K., & Tortajada, C. (2019). Water crisis and water wars: Myths and realities. *International Journal of Water Resources Development, 35*(5), 727–731. https://doi.org/10.1080/07900627.2019.1636502.
13. Bodmer, P., Wilkinson, J., & Lorke, A. (2020). Sediment properties drive spatial variability of potential methane production and oxidation in small streams. *Journal of Geophysical Research: Bio Geosciences, 125*(1), e2019JG005213. https://doi.org/10.1029/2019JG005213.
14. Bodzek, M., Konieczny, K., & Kwiecińska, A. (2011). Application of membrane processes in drinking water treatment–State of art. *Desalination and Water Treatment, 35*(1–3), 164–184. https://doi.org/10.5004/dwt.2011.2435.
15. Bouwman, A. F., Lee, D. S., Asman, W. a. H., Dentener, F. J., Van Der Hoek, K. W., & Olivier, J. G. J. (1997). A global high-resolution emission inventory for ammonia. *Global Biogeochemical Cycles, 11*(4), 561–587. https://doi.org/10.1029/97GB02266.
16. Braida, W. J., & Ong, S. K. (1998). Air sparging: Air-water mass transfer coefficients. *Water Resources Research, 34*(12), 3245–3253. https://doi.org/10.1029/98WR02533.
17. Bridges, T. S., Apitz, S. E., Evison, L., Keckler, K., Logan, M., Nadeau, S., & Wenning, R. J. (2006). Risk-based decision making to manage contaminated sediments. *Integrated Environmental Assessment and Management, 2*(1), 51–58. https://doi.org/10.1002/ieam.5630020110.

18. Brown, T. C., Foti, R., & Ramirez, J. A. (2013). Projected freshwater withdrawals in the United States under a changing climate. *Water Resources Research, 49*(3), 1259–1276. https://doi.org/10.1002/wrcr.20076.

19. Bustillo Revuelta, M. (2024). Geological occurrence. In M. Bustillo Revuelta (Ed.), *The Basics of Aggregates* (pp. 53–100). Springer International Publishing. https://doi.org/10.1007/978-3-031-42961-3_3.

20. Cameselle, C., & Reddy, K. R. (2012). Development and enhancement of electro-osmotic flow for the removal of contaminants from soils. *Electrochimica Acta, 86*, 10–22. https://doi.org/10.1016/j.electacta.2012.06.121.

21. Canfield, D. E., Erik Kristensen, & Bo Thamdrup. (2005). The iron and manganese cycles. In D. E. Canfield, E. Kristensen, & B. Thamdrup (Eds.), *Advances in Marine Biology* (Vol. 48, pp. 269–312). Academic Press. https://doi.org/10.1016/S0065-2881(05)48008-6.

22. Carbonell, G., Ramos, C., Pablos, M. V., Ortiz, J. A., & Tarazona, J. V. (2000). A system dynamic model for the assessment of different exposure routes in aquatic ecosystems. *Science of The Total Environment, 247*(2), 107–118. https://doi.org/10.1016/S0048-9697(99)00482-9.

23. Carstea, E. M., Bridgeman, J., Baker, A., & Reynolds, D. M. (2016). Fluorescence spectroscopy for wastewater monitoring: A review. *Water Research, 95*, 205–219. https://doi.org/10.1016/j.watres.2016.03.021.

24. Casero, I., Sicilia, D., Rubio, S., & Pérez-Bendito, D. (1997). Analytical potential of mixed micelle-based methodology for the determination of ionic surfactants. *Analytica Chimica Acta, 345*(1), 75–86. https://doi.org/10.1016/S0003-2670(97)00081-0.

25. Castro, K., & Zander, A. K. (1995). Membrane air-stripping: Effects of pretreatment. *Journal AWWA, 87*(3), 50–61. https://doi.org/10.1002/j.1551-8833.1995.tb06332.x.

26. Chowdary, V. M., Rao, N. H., & Sarma, P. B. S. (2005). Decision support framework for assessment of non-point-source pollution of groundwater in large irrigation projects. *Agricultural Water Management, 75*(3), 194–225. https://doi.org/10.1016/j.agwat.2004.12.013.

27. Christensen, T. H., Bjerg, P. L., Banwart, S. A., Jakobsen, R., Heron, G., & Albrechtsen, H.-J. (2000). Characterization of redox conditions in groundwater contaminant plumes. *Journal of Contaminant Hydrology, 45*(3), 165–241. https://doi.org/10.1016/S0169-7722(00)00109-1.

28. Ciampi, P., Esposito, C., Bartsch, E., Alesi, E. J., & Petrangeli Papini, M. (2023). Pump-and-treat (P&T) vs groundwater circulation wells (GCW): Which approach delivers more sustainable and effective groundwater

remediation? *Environmental Research, 234*, 116538. https://doi.org/10.1016/j.envres.2023.116538.

29. Colombano, S., Davarzani, H., van Hullebusch, E. D., Ignatiadis, I., Huguenot, D., Omirbekov, S., & Guyonnet, D. (2020). Free product recovery of non-aqueous phase liquids in contaminated sites: theory and case studies. In E. D. van Hullebusch, D. Huguenot, Y. Pechaud, M.-O. Simonnot, & S. Colombano (Eds.), *Environmental Soil Remediation and Rehabilitation: Existing and Innovative Solutions* (pp. 61–148). Springer International Publishing. https://doi.org/10.1007/978-3-030-40348-5_2.
30. Cravotta, C. A. (2008). Dissolved metals and associated constituents in abandoned coal-mine discharges, Pennsylvania, USA. Part 1: Constituent quantities and correlations. *Applied Geochemistry, 23*(2), 166–202. https://doi.org/10.1016/j.apgeochem.2007.10.011.
31. Damalas, C. A., Telidis, G. K., & Thanos, S. D. (2008). Assessing farmers' practices on disposal of pesticide waste after use. *Science of The Total Environment, 390*(2), 341–345. https://doi.org/10.1016/j.scitotenv.2007.10.028.
32. Dearden, R. A., Noy, D. J., Lelliott, M. R., Wilson, R., & Wealthall, G. P. (2013). Release of contaminants from a heterogeneously fractured low permeability unit underlying a DNAPL source zone. *Journal of Contaminant Hydrology, 153*, 141–155. https://doi.org/10.1016/j.jconhyd.2011.05.006.
33. Dhanu Radha, S. V. V., Sabarathinam, C., Al-Ayyadhi, N., Al-Ajeel, F. K., Al-Qallaf, H., & Akber, A. (2022). Spatial and temporal variation of dissolved CO_2 in rainwater from an arid region with special focus on its association with DIC and pCO2. *Environmental Earth Sciences, 81*(4), 113. https://doi.org/10.1007/s12665-022-10176-4.
34. Dutton, A., Wilkinson, B. H., Welker, J. M., Bowen, G. J., & Lohmann, K. C. (2005). Spatial distribution and seasonal variation in 18O/16O of modern precipitation and river water across the conterminous USA. *Hydrological Processes, 19*(20), 4121–4146. https://doi.org/10.1002/hyp.5876.
35. Einsele, G., Yan, J., & Hinderer, M. (2001). Atmospheric carbon burial in modern lake basins and its significance for the global carbon budget. *Global and Planetary Change, 30*(3), 167–195. https://doi.org/10.1016/S0921-8181(01)00105-9.
36. Faimon, J., & Lang, M. (2018). What actually controls the minute to hour changes in soil carbon dioxide concentrations? *Geoderma, 323*, 52–64. https://doi.org/10.1016/j.geoderma.2018.02.048.
37. Fram, M. S., & Belitz, K. (2011). Occurrence and concentrations of

pharmaceutical compounds in groundwater used for public drinking-water supply in California. *Science of The Total Environment, 409*(18), 3409–3417. https://doi.org/10.1016/j.scitotenv.2011.05.053.

38. Fulazzaky, M. A. (2014). Challenges of integrated water resources management in Indonesia. *Water, 6*(7), Article 7. https://doi.org/10.3390/w6072000.
39. Gabr, M. A., Sharmin, N., & Quaranta, J. D. (2013). Multiphase extraction of light non-aqueous phase liquid (lnapl) using prefabricated vertical wells. *Geotechnical and Geological Engineering, 31*(1), 103–118. https://doi.org/10.1007/s10706-012-9567-5.
40. Ganiyu, S. A., Oyadeyi, A. T., & Adeyemi, A. A. (2021). Assessment of heavy metals contamination and associated risks in shallow groundwater sources from three different residential areas within Ibadan metropolis, southwest Nigeria. *Applied Water Science, 11*(5), 81. https://doi.org/10.1007/s13201-021-01414-4.
41. Gantner, S., Andersson, A. F., Alonso-Sáez, L., & Bertilsson, S. (2011). Novel primers for 16S rRNA-based archaeal community analyses in environmental samples. *Journal of Microbiological Methods, 84*(1), 12–18. https://doi.org/10.1016/j.mimet.2010.10.001.
42. Glynn, P. D., & Plummer, L. N. (2005). Geochemistry and the understanding of ground-water systems. *Hydrogeology Journal, 13*(1), 263–287. https://doi.org/10.1007/s10040-004-0429-y.
43. Gomes, C. de S. F., & Silva, J. B. P. (2007). Minerals and clay minerals in medical geology. *Applied Clay Science, 36*(1), 4–21. https://doi.org/10.1016/j.clay.2006.08.006.
44. Grabowski, R. C., Droppo, I. G., & Wharton, G. (2011). Erodibility of cohesive sediment: The importance of sediment properties. *Earth-Science Reviews, 105*(3), 101–120. https://doi.org/10.1016/j.earscirev.2011.01.008.
45. Hanjra, M. A., & Qureshi, M. E. (2010). Global water crisis and future food security in an era of climate change. *Food Policy, 35*(5), 365–377. https://doi.org/10.1016/j.foodpol.2010.05.006.
46. Hardy, S. D., & Koontz, T. M. (2008). Reducing nonpoint source pollution through collaboration: Policies and programs across the U.S. States. *Environmental Management, 41*(3), 301–310. https://doi.org/10.1007/s00267-007-9038-6.
47. Havril, T., Molson, J. W., & Mádl-Szőnyi, J. (2016). Evolution of fluid flow and heat distribution over geological time scales at the margin of unconfined and confined carbonate sequences—A numerical investigation

based on the Buda Thermal Karst analogue. *Marine and Petroleum Geology, 78*, 738–749. https://doi.org/10.1016/j.marpetgeo.2016.10.001.

48. Huang, X., Sillanpää, M., Gjessing, E. T., & Vogt, R. D. (2009). Water quality in the Tibetan plateau: Major ions and trace elements in the headwaters of four major Asian rivers. *Science of The Total Environment, 407*(24), 6242–6254. https://doi.org/10.1016/j.scitotenv.2009.09.001.
49. Hull, R. M., Krogmann, U., & Strom, P. F. (2005). Composition and characteristics of excavated materials from a new jersey landfill. *Journal of Environmental Engineering, 131*(3), 478–490. https://doi.org/10.1061/(ASCE)0733-9372(2005)131:3(478).
50. Ingraham, N. L. (1998). Chapter 3—Isotopic variations in precipitation. In C. Kendall & J. J. McDonnell (Eds.), *Isotope Tracers in Catchment Hydrology* (pp. 87–118). Elsevier. https://doi.org/10.1016/B978-0-444-81546-0.50010-0.
51. Iranpour, R., Cox, H. H. J., Deshusses, M. A., & Schroeder, E. D. (2005). Literature review of air pollution control biofilters and bio trickling filters for odor and volatile organic compound removal. *Environmental Progress, 24*(3), 254–267. https://doi.org/10.1002/ep.10077.
52. Jeon, J.-S., Cho, G., Kim, S., Riu, M., & Song, J. (2024). Actinomycetal, a central constituent microbe during long-term exposure to diazinon, an organophosphorus insecticide. *Chemosphere, 354*, 141583. https://doi.org/10.1016/j.chemosphere.2024.141583.
53. Jones, D. L., Nguyen, C., & Finlay, R. D. (2009). Carbon flow in the rhizosphere: Carbon trading at the soil–Root interface. *Plant and Soil, 321*(1), 5–33. https://doi.org/10.1007/s11104-009-9925-0.
54. Kaiser, K., & Guggenberger, G. (2003). Mineral surfaces and soil organic matter. *European Journal of Soil Science, 54*(2), 219–236. https://doi.org/10.1046/j.1365-2389.2003.00544.x.
55. Karberg, N. J., Pregitzer, K. S., King, J. S., Friend, A. L., & Wood, J. R. (2005). Soil carbon dioxide partial pressure and dissolved inorganic carbonate chemistry under elevated carbon dioxide and ozone. *Oecologia, 142*(2), 296–306. https://doi.org/10.1007/s00442-004-1665-5.
56. Kashyap, S. J., Sankannavar, R., & Madhu, G. M. (2021). Fluoride sources, toxicity and fluorosis management techniques–A brief review. *Journal of Hazardous Materials Letters, 2*, 100033. https://doi.org/10.1016/j.hazl.2021.100033.
57. Khaitan, S., Kalainesan, S., Erickson, L. E., Kulakow, P., Martin, S., Karthikeyan, R., Hutchinson, S. L. L., Davis, L. C., Illangasekare, T. H., & Ng'oma, C. (2006). Remediation of sites contaminated by oil

refinery operations. *Environmental Progress, 25*(1), 20–31. https://doi.org/10.1002/ep.10083.

58. Khangaonkar, T., Sackmann, B., Long, W., Mohamedali, T., & Roberts, M. (2012). Simulation of annual biogeochemical cycles of nutrient balance, phytoplankton bloom(s), and DO in Puget Sound using an unstructured grid model. *Ocean Dynamics, 62*(9), 1353–1379. https://doi.org/10.1007/s10236-012-0562-4.
59. Kim, J., & Choi, H. (2002). Modeling in situ ozonation for the remediation of nonvolatile PAH-contaminated unsaturated soils. *Journal of Contaminant Hydrology, 55*(3), 261–285. https://doi.org/10.1016/S0169-7722(01)00196-6.
60. Kivits, T., Broers, H. P., Beeltje, H., van Vliet, M., & Griffioen, J. (2018). Presence and fate of veterinary antibiotics in age-dated groundwater in areas with intensive livestock farming. *Environmental Pollution, 241*, 988–998. https://doi.org/10.1016/j.envpol.2018.05.085.
61. Kjær, K. H., Bjørk, A. A., Kjeldsen, K. K., Hansen, E. S., Andresen, C. S., Siggaard-Andersen, M.-L., Khan, S. A., Søndergaard, A. S., Colgan, W., Schomacker, A., Woodroffe, S., Funder, S., Rouillard, A., Jensen, J. F., & Larsen, N. K. (2022). Glacier response to the Little Ice Age during the Neoglacial cooling in Greenland. *Earth-Science Reviews, 227*, 103984. https://doi.org/10.1016/j.earscirev.2022.103984.
62. Kladivko, E. J., Brown, L. C., & Baker, J. L. (2001). Pesticide transport to subsurface tile drains in humid regions of North America. *Critical Reviews in Environmental Science and Technology, 31*(1), 1–62. https://doi.org/10.1080/20016491089163.
63. Köhne, J. M., Köhne, S., & Šimůnek, J. (2009). A review of model applications for structured soils: B) Pesticide transport. *Journal of Contaminant Hydrology, 104*(1), 36–60. https://doi.org/10.1016/j.jconhyd.2008.10.003.
64. Krause, L., Klumpp, E., Nofz, I., Missong, A., Amelung, W., & Siebers, N. (2020). Colloidal iron and organic carbon control soil aggregate formation and stability in arable Luvisols. *Geoderma, 374*, 114421. https://doi.org/10.1016/j.geoderma.2020.114421.
65. Kueper, B. H., Stroo, H. F., Vogel, C. M., & Ward, C. H. (Eds.). (2014). *Chlorinated Solvent Source Zone Remediation*. Springer. https://doi.org/10.1007/978-1-4614-6922-3.
66. Kuppusamy, S., Maddela, N. R., Megharaj, M., & Venkateswarlu, K. (2020). Fate of Total Petroleum Hydrocarbons in the Environment. In S. Kuppusamy, N. R. Maddela, M. Megharaj, & K. Venkateswarlu (Eds.), *Total Petroleum Hydrocarbons: Environmental Fate, Toxicity, and*

Remediation (pp. 57–77). Springer International Publishing. https://doi.org/10.1007/978-3-030-24035-6_3.

67. Kwok, C. K., Liang, Y., Leung, S. Y., Wang, H., Dong, Y. H., Young, L., Giesy, J. P., & Wong, M. H. (2013). Biota–Sediment accumulation factor (BSAF), bioaccumulation factor (BAF), and contaminant levels in prey fish to indicate the extent of PAHs and OCPs contamination in eggs of waterbirds. *Environmental Science and Pollution Research, 20*(12), 8425–8434. https://doi.org/10.1007/s11356-013-1809-4.
68. Lin, G.-H., Sauer, N. E., & Cutright, T. J. (1996). Environmental regulations: A brief overview of their applications to bioremediation. *International Biodeterioration & Biodegradation, 38*(1), 1–8. https://doi.org/10.1016/S0964-8305(96)00007-8.
69. Lin, K.-S., Mdlovu, N. V., Chen, C.-Y., Chiang, C.-L., & Dehvari, K. (2018). Degradation of TCE, PCE, and 1,2–DCE DNAPLs in contaminated groundwater using polyethyleneimine-modified zero-valent iron nanoparticles. *Journal of Cleaner Production, 175*, 456–466. https://doi.org/10.1016/j.jclepro.2017.12.074.
70. Linek, V., Sinkule✠, J., & Janda, V. (1998). Design of packed aeration towers to strip volatile organic contaminants from water. *Water Research, 32*(4), 1264–1270. https://doi.org/10.1016/S0043-1354(97)00336-9.
71. Lisitzin, A. P. (2011). Arid sedimentation in the oceans and atmospheric particulate matter. *Russian Geology and Geophysics, 52*(10), 1100–1133. https://doi.org/10.1016/j.rgg.2011.09.006.
72. Liu, Y., Lonappan, L., Brar, S. K., & Yang, S. (2018). Impact of biochar amendment in agricultural soils on the sorption, desorption, and degradation of pesticides: A review. *Science of The Total Environment, 645*, 60–70. https://doi.org/10.1016/j.scitotenv.2018.07.099.
73. Luo, Q., Wu, J., Yang, Y., Qian, J., & Wu, J. (2016). Multi-objective optimization of long-term groundwater monitoring network design using a probabilistic Pareto genetic algorithm under uncertainty. *Journal of Hydrology, 534*, 352–363. https://doi.org/10.1016/j.jhydrol.2016.01.009.
74. Maiphetlho, K., Chimuka, L., Tutu, H., & Richards, H. (2021). Technical design and optimization of polymer inclusion membranes (PIMs) for sample pre-treatment and passive sampling–A review. *Science of The Total Environment, 799*, 149483. https://doi.org/10.1016/j.scitotenv.2021.149483.
75. Maire, J., Joubert, A., Kaifas, D., Invernizzi, T., Marduel, J., Colombano, S., Cazaux, D., Marion, C., Klein, P.-Y., Dumestre, A., & Fatin-Rouge, N. (2018). Assessment of flushing methods for the removal of heavy chlorinated compounds DNAPL in an alluvial aquifer. *Science of*

The Total Environment, 612, 1149–1158. https://doi.org/10.1016/j.scitotenv.2017.08.309.

76. Malik, S., Beer, M., Megharaj, M., & Naidu, R. (2008). The use of molecular techniques to characterize the microbial communities in contaminated soil and water. *Environment International, 34*(2), 265–276. https://doi.org/10.1016/j.envint.2007.09.001.
77. Marchetti, M. (2002). Environmental changes in the central Po Plain (northern Italy) due to fluvial modifications and anthropogenic activities. *Geomorphology, 44*(3), 361–373. https://doi.org/10.1016/S0169-555X(01)00183-0.
78. Martínez-Santos, P. (2017). Does 91% of the world's population really have "sustainable access to safe drinking water"? *International Journal of Water Resources Development, 33*(4), 514–533. https://doi.org/10.1080/07900627.2017.1298517.
79. Marzadri, A., Tonina, D., McKean, J. A., Tiedemann, M. G., & Benjankar, R. M. (2014). Multi-scale streambed topographic and discharge effects on hyporheic exchange at the stream network scale in confined streams. *Journal of Hydrology, 519*, 1997–2011. https://doi.org/10.1016/j.jhydrol.2014.09.076.
80. Matiatos, I., Moeck, C., Vystavna, Y., Marttila, H., Orlowski, N., Jessen, S., Evaristo, J., Sebilo, M., Koren, G., Dimitriou, E., Müller, S., Panagopoulos, Y., & Stockinger, M. P. (2023). Nitrate isotopes in catchment hydrology: Insights, ideas and implications for models. *Journal of Hydrology, 626*, 130326. https://doi.org/10.1016/j.jhydrol.2023.130326.
81. McCarty, P. L. (2010). Groundwater Contamination by Chlorinated Solvents: History, Remediation Technologies and Strategies. In H. F. Stroo & C. H. Ward (Eds.), *In Situ Remediation of Chlorinated Solvent Plumes* (pp. 1–28). Springer. https://doi.org/10.1007/978-1-4419-1401-9_1.
82. McDermott, F. (2004). Palaeo-climate reconstruction from stable isotope variations in speleothems: A review. *Quaternary Science Reviews, 23*(7), 901–918. https://doi.org/10.1016/j.quascirev.2003.06.021.
83. Méité, D., Chesnaux, R., Rafini, S., & Ferroud, A. (2023). Characterizing the radius of influence during pumping tests using the absolute critical drawdown criterion: Cases of integer flow dimensions. *Journal of Hydrology, 618*, 129096. https://doi.org/10.1016/j.jhydrol.2023.129096.
84. Mekala, C., Gaonkar, O., & Nambi, I. M. (2017). Understanding nitrogen and carbon biogeotransformations and transport dynamics in saturated soil columns. *Geoderma, 285*, 185–194. https://doi.org/10.1016/j.geoderma.2016.10.004.

85. Merino, C., Kuzyakov, Y., Godoy, K., Jofré, I., Nájera, F., & Matus, F. (2021). Iron-reducing bacteria decompose lignin by electron transfer from soil organic matter. *Science of The Total Environment, 761*, 143194. https://doi.org/10.1016/j.scitotenv.2020.143194.

86. Morillo, E., & Villaverde, J. (2017). Advanced technologies for the remediation of pesticide-contaminated soils. *Science of The Total Environment, 586*, 576–597. https://doi.org/10.1016/j.scitotenv.2017.02.020.

87. Mu, Y., Yuan, Y., Jia, X., Zha, T., Qin, S., Ye, Z., Liu, P., Yang, R., & Tian, Y. (2022). Hydrological losses and soil moisture carryover affected the relationship between evapotranspiration and rainfall in a temperate semiarid shrubland. *Agricultural and Forest Meteorology, 315*, 108831. https://doi.org/10.1016/j.agrformet.2022.108831.

88. Mulligan, A. E., & Ahlfeld, D. P. (1999). Advective control of groundwater contaminant plumes: Model development and comparison to hydraulic control. *Water Resources Research, 35*(8), 2285–2294. https://doi.org/10.1029/1999WR900106.

89. Munksgaard, N. C., & Nelson, P. N. (2021). Coupled polymer-membrane equilibration and cavity ring-down spectrometry for the highly sensitive determination of dissolved methane in environmental waters. *Analytical Letters, 54*(3), 430–441. https://doi.org/10.1080/00032719.2020.1767122
.

90. Naidu, R., Sumner, M. E., & Harter, R. D. (1998). Sorption of heavy metals in strongly weathered soils: An overview. *Environmental Geochemistry and Health, 20*(1), 5–9. https://doi.org/10.1023/A:1006519009465.

91. Narasimhan, T. n. (1998). Hydraulic characterization of aquifers, reservoir rocks, and soils: A history of ideas. *Water Resources Research, 34*(1), 33–46. https://doi.org/10.1029/97WR02714.

92. Naveen, B. P., Mahapatra, D. M., Sitharam, T. G., Sivapullaiah, P. V., & Ramachandra, T. V. (2017). Physico-chemical and biological characterization of urban municipal landfill leachate. *Environmental Pollution, 220*, 1–12. https://doi.org/10.1016/j.envpol.2016.09.002.

93. Nizoli, E. C., & Luiz-Silva, W. (2012). Seasonal AVS–SEM relationship in sediments and potential bioavailability of metals in industrialized estuary, southeastern Brazil. *Environmental Geochemistry and Health, 34*(2), 263–272. https://doi.org/10.1007/s10653-011-9430-2.

94. Obiri-Nyarko, F., Grajales-Mesa, S. J., & Malina, G. (2014). An overview of permeable reactive barriers for in situ sustainable groundwater remediation. *Chemosphere, 111*, 243–259. https://doi.org/10.1016/j.chemosphere.2014.03.112.

95. Olutona, G. O., Olatunji, S. O., & Obisanya, J. F. (2016). Downstream assessment of chlorinated organic compounds in the bed-sediment of Abia Stream, Iwo, South-Western, Nigeria. *Springer Plus, 5*(1), 67. https://doi.org/10.1186/s40064-016-1664-0.

96. Ongley, E. D., Krishnappan, B. G., Droppo, G., Rao, S. S., & Maguire, R. J. (1992). Cohesive sediment transport: Emerging issues for toxic chemical management. *Hydrobiologia, 235*(1), 177–187. https://doi.org/10.1007/BF00026210.

97. Or, D., Smets, B. F., Wraith, J. M., Dechesne, A., & Friedman, S. P. (2007). Physical constraints affecting bacterial habitats and activity in unsaturated porous media–A review. *Advances in Water Resources, 30*(6), 1505–1527. https://doi.org/10.1016/j.advwatres.2006.05.025.

98. Pan, D., Fan, Q., Li, P., Liu, S., & Wu, W. (2011). Sorption of Th (IV) on Na-bentonite: Effects of pH, ionic strength, humic substances and temperature. *Chemical Engineering Journal, 172*(2), 898–905. https://doi.org/10.1016/j.cej.2011.06.080.

99. Paszko, T., Muszyński, P., Materska, M., Bojanowska, M., Kostecka, M., & Jackowska, I. (2016). Adsorption and degradation of phenoxy alkanoic acid herbicides in soils: A review. *Environmental Toxicology and Chemistry, 35*(2), 271–286. https://doi.org/10.1002/etc.3212.

100. Pfister, S., Bayer, P., Koehler, A., & Hellweg, S. (2011). Projected water consumption in future global agriculture: Scenarios and related impacts. *Science of The Total Environment, 409*(20), 4206–4216. https://doi.org/10.1016/j.scitotenv.2011.07.019.

101. Poeplau, C., Don, A., & Schneider, F. (2021). Roots are key to increasing the mean residence time of organic carbon entering temperate agricultural soils. *Global Change Biology, 27*(19), 4921–4934. https://doi.org/10.1111/gcb.15787.

102. Postigo, C., & Barceló, D. (2015). Synthetic organic compounds and their transformation products in groundwater: Occurrence, fate and mitigation. *Science of The Total Environment, 503–504*, 32–47. https://doi.org/10.1016/j.scitotenv.2014.06.019.

103. Quiñonez-Plaza, A., Wakida, F. T., Temores-Peña, J., Rodriguez-Mendivil, D. D., Garcia-Flores, E., Pastrana-Corral, M. A., & Melendez-Lopez, S. G. (2017). Total petroleum hydrocarbons and heavy metals in road-deposited sediments in Tijuana, Mexico. *Journal of Soils and Sediments, 17*(12), 2873–2886. https://doi.org/10.1007/s11368-017-1778-1.

104. Raessler, M. (2018). The arsenic contamination of drinking and groundwaters in Bangladesh: Featuring biogeochemical aspects and implications on public health. *Archives of Environmental Contamination*

and Toxicology, 75(1), 1–7. https://doi.org/10.1007/s00244-018-0511-4.

105. Reed, P. M., Hadka, D., Herman, J. D., Kasprzyk, J. R., & Kollat, J. B. (2013). Evolutionary mult objective optimization in water resources: The past, present, and future. *Advances in Water Resources, 51*, 438–456. https://doi.org/10.1016/j.advwatres.2012.01.005.
106. Rivett, M. O., Wealthall, G. P., Dearden, R. A., & McAlary, T. A. (2011). Review of unsaturated-zone transport and attenuation of volatile organic compound (VOC) plumes leached from shallow source zones. *Journal of Contaminant Hydrology, 123*(3), 130–156. https://doi.org/10.1016/j.jconhyd.2010.12.013.
107. Rodhe, A., & Killingtveit, Â. (1996). Catchment hydrology. In *Geochemical Processes, Weathering and Groundwater Recharge in Catchments*. CRC Press.
108. Rozzi, A., & Remigi, E. (2004). Methods of assessing microbial activity and inhibition under anaerobic conditions: A literature review. *Reviews in Environmental Science & Biotechnology, 3*(2), 93–115. https://doi.org/10.1007/s11157-004-5762-z.
109. Rupias, O. J. B., Pereira, S. Y., & de Abreu, A. E. S. (2021). Hydro geochemistry and groundwater quality assessment using the water quality index and heavy-metal pollution index in the alluvial plain of Atibaia river–Campinas/SP, Brazil. *Groundwater for Sustainable Development, 15*, 100661. https://doi.org/10.1016/j.gsd.2021.100661.
110. Safeeq, M., & Fares, A. (2016). Groundwater and surface water interactions in relation to natural and anthropogenic environmental changes. In A. Fares (Ed.), *Emerging Issues in Groundwater Resources* (pp. 289–326). Springer International Publishing. https://doi.org/10.1007/978-3-319-32008-3_11.
111. Samanta, S., Dalai, T. K., Pattanaik, J. K., Rai, S. K., & Mazumdar, A. (2015). Dissolved inorganic carbon (DIC) and its δ13C in the Ganga (Hooghly) River estuary, India: Evidence of DIC generation via organic carbon degradation and carbonate dissolution. *Geochimica et Cosmochimica Acta, 165*, 226–248. https://doi.org/10.1016/j.gca.2015.05.040.
112. Sanchez-Vila, X., Guadagnini, A., & Carrera, J. (2006). Representative hydraulic conductivities in saturated groundwater flow. *Reviews of Geophysics, 44*(3). https://doi.org/10.1029/2005RG000169.
113. Saravanan, A., Senthil Kumar, P., Jeevanantham, S., Karishma, S., Tajsabreen, B., Yaashikaa, P. R., & Reshma, B. (2021). Effective water/wastewater treatment methodologies for toxic pollutants removal: Processes and applications towards sustainable development. *Chemosphere, 280*, 130595. https://doi.org/10.1016/j.chemosphere.2021.130595.

114. Schott, J., Mavromatis, V., Fujii, T., Pearce, C. R., & Oelkers, E. H. (2016). The control of carbonate mineral Mg isotope composition by aqueous speciation: Theoretical and experimental modeling. *Chemical Geology, 445*, 120–134. https://doi.org/10.1016/j.chemgeo.2016.03.011.

115. Schulze, F., Gao, X., Virzonis, D., Damiati, S., Schneider, M. R., & Kodzius, R. (2017). Air quality effects on human health and approaches for its assessment through microfluidic chips. *Genes, 8*(10), Article 244. https://doi.org/10.3390/genes8100244.

116. Singh, A. K., Varma, N. P., & Mondal, G. C. (2016). Hydrogeochemical investigation and quality assessment of mine water resources in the Korba coalfield, India. *Arabian Journal of Geosciences, 9*(4), 278. https://doi.org/10.1007/s12517-015-2298-1.

117. Sophocleous, M. (2002). Interactions between groundwater and surface water: The state of the science. *Hydrogeology Journal, 10*(1), 52–67. https://doi.org/10.1007/s10040-001-0170-8.

118. Spencer, W. F., & Cliath, M. M. (1973). Pesticide volatilization as related to water loss from soil. *Journal of Environmental Quality, 2*(2), 284–289. https://doi.org/10.2134/jeq1973.00472425000200020026x.

119. Srivastav, A. L. (2020). Chapter 6–Chemical fertilizers and pesticides: Role in groundwater contamination. In M. N. V. Prasad (Ed.), *Agrochemicals Detection, Treatment and Remediation* (pp. 143–159). Butterworth-Heinemann. https://doi.org/10.1016/B978-0-08-103017-2.00006-4.

120. Stadler, H., Leis, A., Plieschnegger, M., Skritek, P., Woletz, K., & Farnleitner, A. (2011). Automated event sampling and real-time data access within hydrological measuring networks by means of low-earth-orbiting satellites. In P. Golinska, M. Fertsch, & J. Marx-Gómez (Eds.), *Information Technologies in Environmental Engineering: New Trends and Challenges* (pp. 725–734). Springer. https://doi.org/10.1007/978-3-642-19536-5_56.

121. Starr, C. (1993). Atmospheric CO2 residence time and the carbon cycle. *Energy, 18*(12), 1297–1310. https://doi.org/10.1016/0360-5442(93)90017-8.

122. Suthersan, S. S., Horst, J., Schnobrich, M., Welty, N., & McDonough, J. (2016). *Remediation Engineering: Design Concepts, Second Edition* (2nd ed.). CRC Press. https://doi.org/10.1201/9781315367088.

123. Theodore, M. K., & Theodore, L. (2021). Industrial wastewater management. In *Introduction to Environmental Management* (2nd ed.). CRC Press.

124. Thévenot, D. R., Moilleron, R., Lestel, L., Gromaire, M.-C., Rocher,

V., Cambier, P., Bonté, P., Colin, J.-L., de Pontevès, C., & Meybeck, M. (2007). Critical budget of metal sources and pathways in the Seine River basin (1994–2003) for Cd, Cr, Cu, Hg, Ni, Pb and Zn. *Science of The Total Environment, 375*(1), 180–203. https://doi.org/10.1016/j.scitotenv.2006.12.008.

125. Thompson, J. M. (1979). Arsenic and fluoride in the upper madison river system: Firehole and gibbon rivers and their tributaries, Yellowstone national park, Wyoming, and southeast Montana. *Environmental Geology, 3*(1), 13–21. https://doi.org/10.1007/BF02423274.
126. Valavanidis, A., Fiotakis, K., & Vlachogianni, T. (2008). Airborne particulate matter and human health: toxicological assessment and importance of size and composition of particles for oxidative damage and carcinogenic mechanisms. *Journal of Environmental Science and Health, Part C, 26*(4), 339–362. https://doi.org/10.1080/10590500802494538.
127. Vorkamp, K., Herrmann, R., & Hvitved-Jacobsen, T. (2001). Characterization of organic matter from anaerobic digestion of organic waste by aerobic microbial activity. *Bioresource Technology, 78*(3), 257–265. https://doi.org/10.1016/S0960-8524(01)00021-9.
128. Wen, Z., Shang, Y., Lyu, L., Tao, H., Liu, G., Fang, C., Li, S., & Song, K. (2024). Re-estimating China's lake CO2 flux considering spatiotemporal variability. *Environmental Science and Ecotechnology, 19*, 100337. https://doi.org/10.1016/j.ese.2023.100337.
129. Wolkoff, P., Schneider, T., Kildesø, J., Degerth, R., Jaroszewski, M., & Schunk, H. (1998). Risk in cleaning: Chemical and physical exposure. *Science of The Total Environment, 215*(1), 135–156. https://doi.org/10.1016/S0048-9697(98)00110-7.
130. Wu, H., He, Y., Bao, L., Zhu, C., & Li, S. (2017). Mineral composition control on inter-mineral iron isotopic fractionation in granitoids. *Geochimica et Cosmochimica Acta, 198*, 208–217. https://doi.org/10.1016/j.gca.2016.11.008.
131. Yan, Z., Wu, J., Wang, X., & Zhang, Y. (2015). Development of water quality criteria for toxic organic pollutants. In Z. Yan & Z. Liu (Eds.), *Toxic Pollutants in China: Study of Water Quality Criteria* (pp. 1–55). Springer Netherlands. https://doi.org/10.1007/978-94-017-9795-5_1.
132. Yang, C., Offiong, N.-A., Zhang, C., Liu, F., & Dong, J. (2021). Mechanisms of irreversible density modification using colloidal biliquid apron for dense nonaqueous phase liquids in contaminated aquifer remediation. *Journal of Hazardous Materials, 415*, 125667. https://doi.org/10.1016/j.jhazmat.2021.125667.
133. Yoon, H., Jun, S.-C., Hyun, Y., Bae, G.-O., & Lee, K.-K. (2011). A

comparative study of artificial neural networks and support vector machines for predicting groundwater levels in a coastal aquifer. *Journal of Hydrology, 396*(1), 128–138. https://doi.org/10.1016/j.jhydrol.2010.11.002.

134. Zeinalie, M., Bozorg-Haddad, O., & Chaplot, B. (2021). Modeling and Simulation in Water Resources Management. In O. Bozorg-Haddad (Ed.), *Essential Tools for Water Resources Analysis, Planning, and Management* (pp. 1–31). Springer. https://doi.org/10.1007/978-981-33-4295-8_1.
135. Zeng, Q., Huang, H., Tan, Y., Chen, G., & Hao, T. (2022). Emerging electrochemistry-based process for sludge treatment and resources recovery: A review. *Water Research, 209*, 117939. https://doi.org/10.1016/j.watres.2021.117939.
136. Zhang, W. L., Tian, Z. X., Zhang, N., & Li, X. Q. (1996). Nitrate pollution of groundwater in northern China. *Agriculture, Ecosystems & Environment, 59*(3), 223–231. https://doi.org/10.1016/0167-8809(96)01052-3.

INDEX

T

U

V

W

Z

FSC
www.fsc.org
MIX
Papier | Fördert gute Waldnutzung
FSC® C083411

Kolibri 360 GmbH
Ferdinand-Jühlke-Straße 7
99095 Erfurt, Deutschland
produktsicherheit@kolibri360.de